FISH QUALITY CONTROL BY COMPUTER VISION

FOOD SCIENCE AND TECHNOLOGY

A Series of Monographs, Textbooks, and Reference Books

1. Flavor Research: Principles and Techniques, *R. Teranishi, I. Hornstein, P. Issenberg, and E. L. Wick (out of print)*
2. Principles of Enzymology for the Food Sciences, *John R. Whitaker*
3. Low-Temperature Preservation of Foods and Living Matter, *Owen R. Fennema, William D. Powrie, and Elmer H. Marth*
4. Principles of Food Science
   Part I: Food Chemistry, *edited by Owen R. Fennema*
   Part II: Physical Methods of Food Preservation, *Marcus Karel, Owen R. Fennema, and Daryl B. Lund*
5. Food Emulsions, *edited by Stig Friberg*
6. Nutritional and Safety Aspects of Food Processing, *edited by Steven R. Tannenbaum*
7. Flavor Research: Recent Advances, *edited by R. Teranishi, Robert A. Flath, and Hiroshi Sugisawa*
8. Computer-Aided Techniques in Food Technology, *edited by Israel Saguy*
9. Handbook of Tropical Foods, *edited by Harvey T. Chan*
10. Antimicrobials in Foods, *edited by Alfred Larry Branen and P. Michael Davidson*
11. Food Constituents and Food Residues: Their Chromatographic Determination, *edited by James F. Lawrence*
12. Aspartame: Physiology and Biochemistry, *edited by Lewis D. Stegink and L. J. Filer, Jr.*
13. Handbook of Vitamins: Nutritional, Biochemical, and Clinical Aspects, *edited by Lawrence J. Machlin*

14. Starch Conversion Technology, *edited by G. M. A. van Beynum and J. A. Roels*
15. Food Chemistry: Second Edition, Revised and Expanded, *edited by Owen R. Fennema*
16. Sensory Evaluation of Food: Statistical Methods and Procedures, *Michael O'Mahony*
17. Alternative Sweeteners, *edited by Lyn O'Brien Nabors and Robert C. Gelardi*
18. Citrus Fruits and Their Products: Analysis and Technology, *S. V. Ting and Russell L. Rouseff*
19. Engineering Properties of Foods, *edited by M. A. Rao and S. S. H. Rizvi*
20. Umami: A Basic Taste, *edited by Yojiro Kawamura and Morley R. Kare*
21. Food Biotechnology, *edited by Dietrich Knorr*
22. Food Texture: Instrumental and Sensory Measurement, *edited by Howard R. Moskowitz*
23. Seafoods and Fish Oils in Human Health and Disease, *John E. Kinsella*
24. Postharvest Physiology of Vegetables, *edited by J. Weichmann*
25. Handbook of Dietary Fiber: An Applied Approach, *Mark L. Dreher*
26. Food Toxicology, Parts A and B, *Jose M. Concon*
27. Modern Carbohydrate Chemistry, *Roger W. Binkley*
28. Trace Minerals in Foods, *edited by Kenneth T. Smith*
29. Protein Quality and the Effects of Processing, *edited by R. Dixon Phillips and John W. Finley*
30. Adulteration of Fruit Juice Beverages, *edited by Steven Nagy, John A. Attaway, and Martha E. Rhodes*
31. Foodborne Bacterial Pathogens, *edited by Michael P. Doyle*
32. Legumes: Chemistry, Technology, and Human Nutrition, *edited by Ruth H. Matthews*
33. Industrialization of Indigenous Fermented Foods, *edited by Keith H. Steinkraus*
34. International Food Regulation Handbook: Policy • Science • Law, *edited by Roger D. Middlekauff and Philippe Shubik*
35. Food Additives, *edited by A. Larry Branen, P. Michael Davidson, and Seppo Salminen*
36. Safety of Irradiated Foods, *J. F. Diehl*
37. Omega-3 Fatty Acids in Health and Disease, *edited by Robert S. Lees and Marcus Karel*
38. Food Emulsions, Second Edition, Revised and Expanded, *edited by Kåre Larsson and Stig E. Friberg*
39. Seafood: Effects of Technology on Nutrition, *George M. Pigott and Barbee W. Tucker*
40. Handbook of Vitamins, Second Edition, Revised and Expanded, *edited by Lawrence J. Machlin*

41. Handbook of Cereal Science and Technology, *Klaus J. Lorenz and Karel Kulp*
42. Food Processing Operations and Scale-Up, *Kenneth J. Valentas, Leon Levine, and J. Peter Clark*
43. Fish Quality Control by Computer Vision, *edited by L. F. Pau and R. Olafsson*

Other Volumes in Preparation

Volatile Compounds in Foods and Beverages, *edited by H. Maarse*

Listeria, Listeriosis, and Food Safety, *Elliot T. Ryser and Elmer H. Marth*

FISH QUALITY CONTROL BY COMPUTER VISION

edited by

L. F. Pau
Digital Equipment Corporation
Valbonne, France

R. Olafsson
University of Iceland
Reykjavik, Iceland

Marcel Dekker, Inc. New York • Basel • Hong Kong

Library of Congress Cataloging-in-Publication Data

Fish quality control by computer vision / edited by L. F. Pau, R. Olafsson.
p. cm. -- (Food science and technology ; 43)
Includes bibliographical references and index.
ISBN 0-8247-8426-X
1. Fishery products--Quality control. 2. Fishery processing--Quality control. 3. Computer vision. 4. Fish inspection--Quality control. I. Pau, L. F. (Lewes F.) II. R. Olafsson. III. Series.
SH335.5.Q35F57 1991
664'.94--dc20 90-23462
CIP

This book is printed on acid-free paper.

Marcel Dekker, Inc.
270 Madison Avenue, New York, New York 10016

Current printing (last digit):
10 9 8 7 6 5 4 3 2 1

PRINTED IN THE UNITED STATES OF AMERICA

To my daughter Isabelle M. Pau.

L. F. Pau

Preface

The fishing and fish product industries today provide the world's population with a very significant fraction of their nutritional needs. At the same time, ever-stricter health standards are enforced on these products, while at the same time high demands are put on the quality of the raw and processed material. Quite differently from the meat industry, the fishing and fish product industries face a tremendous diversity of catching methods, species, sizes, shapes, and colors, as well as of preparation methods (frozen whole fish, salted whole fish, fillets, fish cubes, etc.). This is why human labor (on board vessels, on board or ashore in freezing plants, processing plants, in fish farms) represents by far the largest cost and also skill factor. Only computer vision, when feasible, can ensure less fatigue, consistent accuracy, higher speed and adjustability, besides offering process-related feedback among handling and processing machines. This involves putting together an application-specific combination of camera sensors, lighting, mechanical handling/reject, real-time digital image processors, algorithms, software, and user interfaces.

The main applications are: 1) quality inspection of fish and fish products for defects, 2) measurement and sorting by length, weight, species, shape, orientation, etc., for packaging, handling, selection, registration, and pricing, 3) analysis for cutting, fileting, etc., 4) analysis of dynamic and age-dependent behavior, and 5) spotting for fish catches.

The structure of this book aims at giving practical solution approaches to the five problem areas above. Chapters 1 and 2 survey the quality assurance requirements and the corresponding operational requirements (speed, accuracy) for the specification of the vision systems. Chapters 2, 3, and 4 emphasize, in relation to biological and optical properties of the fish, the kind of signals or features that allow for the detection of surface defects, as well as of bones and worms. To view these features, the fish or fish products must be suitably illuminated; the choice of the lighting turns out to be very tricky because of varying shapes, specular reflection from water, skin color, etc.; this field is the topic of Chapters 5, 6, and 7. Chapter 8 reviews the state of the art, and some results, in terms of spotting fish banks from satellite imagery, by locating zones where the biological, temperature, and other conditions maximize the likelihood for their presence.

The remaining chapters give solutions to analysis problems from the vision engineering point of view. Chapter 9 gives an architectural overview, while Chapters 10, 11, and 12 describe specialized implementations. Chapter 12 also serves as a hands-on tutorial. Finally, Chapters 13 and 14 give advanced research topics, such as nonimaging sensors for sorting and analysis sensor fusion for inspection and color spectrum illumination for grading.

This volume is the result of a cooperative project undertaken in 1988 and 1989 by about 15 industrial researchers from the Nordic countries (Denmark, Iceland, Norway, Faeroe Islands) and a dozen lecturers and research assistants from Iceland, Denmark, Norway, Finland, France, the United States, and the United Kingdom.

This project, funded by the Nordic Industrial Fund, Oslo, was organized to offer hands-on training and carry out early systems feasibility in the area of computer vision applied to the fishing and fish product industries. It was initiated by the two editors of this volume during a working conference in February 1988 in Copenhagen, on the theme of "Information Processing in the Fish Industry." During the first phase of the work, image acquisition and lighting selection were carried out in Iceland: in fish factories, on a fishing trawler, as well as in the laboratory. During the second phase, these images were processed and algorithms developed for three possible dedicated systems, using four image processors available at the artificial intelligence and vision group of the Technical University of Denmark, Lyngby. During both phases, specialized lectures were given, most of which are included in this volume (after revisions). We are grateful to the authors, as well as to the research assistants and project participants, for their enthusiasm and commitment to a new technology of major importance to the fishing industry. Thanks are also extended to the Nordic Industry Fund, as well as to suppliers of fish samples (in Iceland) and of equipment (AGEMA, Sweden; research establishment of the Danish Ministry of Fisheries), for their material support.

Finally, acknowledgments are extended to Mrs. Claude Mona for help in preparing this manuscript.

L. F. PAU
R. OLAFSSON

Contents

Preface *v*

Contributors *xi*

I. Introduction: Computer Vision Opportunities for the Fish Industry

1. Quality Assurance in the Fishing Industry with Emphasis on the Future Use of Vision Techniques *Jette Nielsen, Jesper Heldbo Reines, and Claus Mosby Jespersen* 3
2. Vision Applications in the Fishing and Fish Product Industries *Hördur Arnarson, Kepa Bengoetxea, and L. F. Pau* 21

II. Biological and Optical Properties of Fish for Vision Applications

3. Optical Spectra of Fish Flesh and Quality Defects in Fish *Jón Pétursson* 45
4. Optical Properties of Fish Classes *Kepa Bengoetxea* 71
5. Parasitic Nematodes in Commercially Important Fish *Erlingur Hauksson* 77

III. Lighting and Vision Techniques

6. Evaluation of Illumination Methods for Machine Vision Applications in the Fish Industry *Timo Piironen* 97
7. Lighting Setup in the Automatic Detection of Ventral Skin and Blood Spots in Cod Fish Fillets *Kepa Bengoetxea* 121
8. Illumination Techniques for 3–D Machine Vision *Alastair McLeod* 151

IV. Remote Sensing

9. Fishery Applications of Satellite Remote Sensing and Image Processing in the Gulf of Mexico *Rex C. Herron* 185

V. Computer Vision Technology for the Fish Industry

10. Image Analysis in the Fish and Food Industries *Norval J. C. Strachan and C. K. Murray* 209
11. Laboratory Work Using the DT 2851 Frame Grabber: Application of Computer Vision to Inspection of Fish *Eric Bardin* 225
12. Fish and Fish Product Sorting *Hördur Arnarson* 245
13. Preliminary Results of a Computer Imaging System Applied to Estimating the Quantity of Larvae and Fingerling Fish for Aquaculture *Michael D. Spratt* 263
14. Advanced Vision Methods and Technologies for the Fishing and Fish Product Industries *L. F. Pau* 283

Appendix: Users of Vision Technology in the Fishing and Fish Processing Industries *291*

Index *293*

Contributors

Hördur Arnarson Department of Research, Marel HF, Reykjavik, Iceland

Eric Bardin AI & Vision Group, EMI, Technical University of Denmark, Lyngby, Denmark

Kepa Bengoetxea Production Department, Dow Chemical Company, Tarragona, Spain

Erlingur Hauksson Research Committee for Biological Seafood Quality, Icelandic Fisheries Laboratories, Reykjavik, Iceland

Rex C. Herron Mississippi Laboratories, Stennis Space Center, National Marine Fisheries Service, Bay St. Louis, Mississippi

Claus Mosby Jespersen Technological Laboratory, Technical University of Denmark, Danish Ministries of Fisheries, Lyngby, Denmark

Alastair McLeod Department of Production Automation, Center for Industrial Research, Oslo, Norway

C. K. Murray Torry Research Station, Ministry of Agriculture, Fisheries and Food, Aberdeen, Scotland

Jette Nielsen Technological Laboratory, Technical University of Denmark, Danish Ministries of Fisheries, Lyngby, Denmark

L. F. Pau* AI & Vision Group, EMI, Technical University of Denmark, Lyngby, Denmark

Jón Pétursson Department of Physics, Science Institute, University of Iceland, Reykjavik, Iceland

Timo Piironen Electronics Laboratory, Technical Research Center of Finland, Oulu, Finland

Jesper Heldbo Reines Technological Laboratory, Technical University of Denmark, Danish Ministries of Fisheries, Lyngby, Denmark

Michael D. Spratt Applying Evolving Technologies, Montpellier, France

Norval J. C. Strachan Torry Research Station, Ministry of Agriculture, Fisheries and Food, Aberdeen, Scotland

**Present affiliation*: AI Group Technical Center, Digital Equipment Corporation, Valbonne, France

I

Introduction: Computer Vision Opportunities for the Fish Industry

1

Quality Assurance in the Fishing Industry with Emphasis on the Future Use of Vision Techniques

JETTE NIELSEN, JESPER HELDBO REINES, and CLAUS MOSBY JESPERSEN *Technical University of Denmark, Lyngby, Denmark*

1. Introduction
2. Quality Control—Quality Assurance
3. The Future of Quality Assurance at Sea
4. Quality Assurance in the Filleting Industry
   4.1 Grading
   4.2 The Fish-Processing Plant
5. Test Systems for Fish Quality
   5.1 Predictive Methods
   5.2 In-Line Methods
   5.3 Control Methods
6. Summary
   References

1. INTRODUCTION

It is necessary to develop principles and ideas of a total quality assurance system covering fish handling at sea and in the fish-processing industry, to secure an increased proportion of high-quality fish resulting in an increased consumption. In the future, big companies and customers will demand that every batch of fish be marked with an information label indicating catch area, time and temperature strain put on the fish, size, and weight. The label, which would guarantee the quality, will be the sales argument of the future for fish. Compared to meat, fish is a very sensitive raw material. The quality losses due to spoilage begin as soon as the fish dies. The mechanism of spoilage is very complex, but the main cause is microbial action or enzymatic/chemical activity. The spoilage may be delayed or prevented by proper handling under a combination of controlled time and temperature; this is very important to take into consideration when designing a quality assurance system.

2. QUALITY CONTROL—QUALITY ASSURANCE

In view of health and hygiene aspects, there exists, at the moment, a rather comprehensive quality control program in the industry; it also covers the inspection of sensory attributes such as taste/odor, texture, and appearance. However, a quality assurance system with a kind of built-in guarantee for high quality will minimize the need for control.

Quality is traditionally considered to be a factor related to microbiological counts and sensory attributes. But quality involves a lot of other factors (the fish shall be easy to handle—technological quality) and can more generally be defined as the fulfillment of the customers' expectations.

Quality assurance in manufacturing refers to the achievement of the quality standards set by the design criteria.

The production manager of a processing plant is faced with a difficult decision, because he never knows exactly what raw material he is going to receive that day. However, by combining the information from a quality assurance program giving reliable information with sales planning, it is possible to give the production manager all the information needed for planning rationally the daily production (11).

Integrated quality assurance is highly dependent on the physical conditions on board fishing vessels, and on the processing as well as the distribution chain. The system must be designed so that the most natural and ergonomical correct production procedure gives the best quality. It has been stated that any attempt to raise the quality of fish is critically dependent on a consistent, easy-to-use method for quality assessment of raw fish. This is not the whole truth, as quality is based on a number of other factors, including

- A rational, mechanized, and ergonomically correct handling system for fish at sea and in the industry ashore
- A mechanized grading system
- A price differential system, which gives a higher price for higher quality
- A method for registration and documentation of the fish which can be used for production planning
- Predictive methods

It is, however, correct that quality assessment of the raw fish is of great importance, as the quality of the product is very seldom better than that of the raw fish.

3. THE FUTURE OF QUALITY ASSURANCE AT SEA

In order to describe a future quality assurance system (Fig. 1) for the industry, it is necessary to predict the development in the immediate future (5 years) for fish handling at sea.

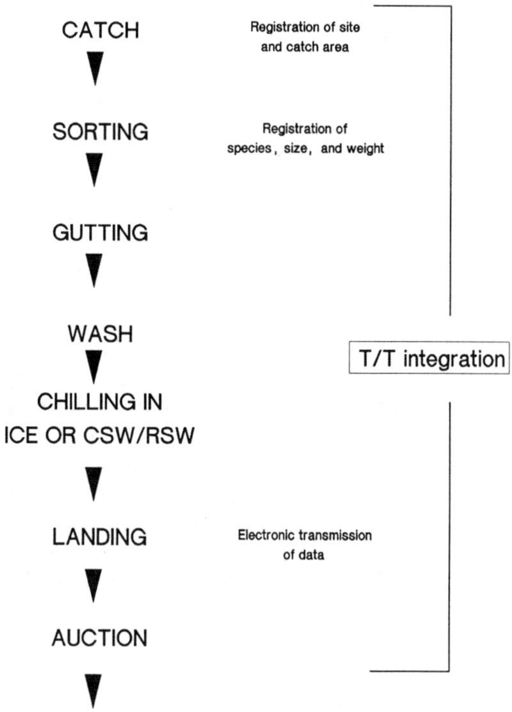

Figure 1 Future system for quality assurance at sea.

Time and temperature are the dominant factors for maintaining the high quality of freshly caught fish. The loss of quality is proportional to time under the same temperature conditions. The optimum storage temperature for fish is between −1 and 0°C, and a temperature increase to just 5°C more than halves the shelf life. This implies that a catch-handling system must be fast and, therefore, as mechanized as possible.

The low temperature must be reached as quickly as possible after catch. This could be reached by a homogeneous mixing of fish and ice and subsequent storage in plastic boxes or by using chilled sea water (CSW) or refrigerated sea water (RSW) tanks for on-board storage. Other factors of importance are bleeding and gutting, as well as the mechanical influence of handling and sorting. The layout of a catch-handling system, including the above

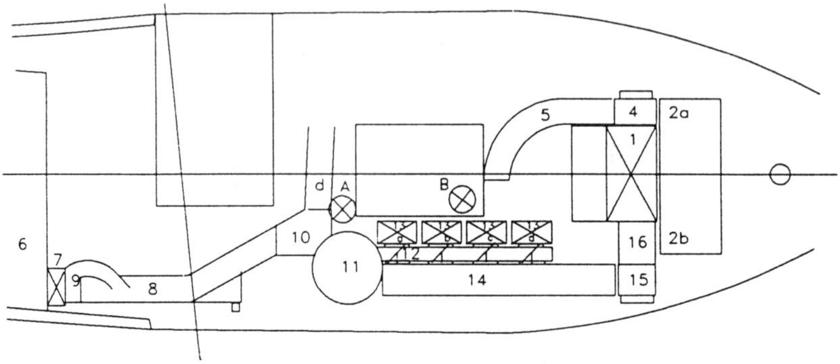

Figure 2 Layout for a 30-m fish trawler with automatic transport of boxes and registration of species, length, and weight. (1) Box elevator for empty/filled boxes from/to fish-hold; (2a & b) ice-silo; (4) tare-scale with bar-code scanner; (5) conveyor for empty boxes; (6) receiving-pounder; (7) elevator for fish; (8) conveyor for fish; (9) skids; (10) buffer-pounder; (11) gutting machine supplied with length measurement device; (12) length-grading device; (13) filling-station for boxes with bar-codes, receiving length-graded fish; (14) conveyor for filled boxes; (15) scale for gross weight; (16) conveyor for filled boxes leading to box elevator. (A & B) Operator.

factors, is therefore of importance. A proposed modern layout is described in Figure 2 (12). Grading of the fish for further processing and packing in either boxes or containers are minimum demands.

The catch-handling system must have a built-in registration and documentation (informative label) for every batch of fish. The informative label shall contain information on time/temperature, species, weight, and length besides catching area and trawling time; this information shall be used when the fish are landed either directly at a processing plant or at a fish auction. On the basis of this system it might be possible in the future to give the fishermen a "just" (fair) price for the fish by using a combination of the obtainable yield and the time/temperature integration.

The amount of meat on a fish is dependent on the biological condition of the fish. The biological condition has a seasonal variation; e.g., the fish contains less flesh around the spawning season. Biologists often measure the length of the fish for later calculation of the weight according to the length/weight relationship for actual species.

The relation is as follows:

$$W = \alpha \times L^{\beta}$$

where:

W = the weight of the fish in grams (eviscerated, head on, gills retained)
L = the total length in centimeters
α = the condition factor
β = the isometric growth exponent

Biologists use the length/weight relationship to estimate the growth and further for stock assessments.

As β is often close to 3 for codfish, α is the amount of flesh on a fish having a certain length in the converse relation:

$$\alpha = \frac{W}{L^{\beta}}$$

and might be used as a predictive measurement of yield.

Decisions for storage, processing, distribution, etc. should be based on the informative label, including the prediction for yield (7).

4. QUALITY ASSURANCE IN THE FILLETING INDUSTRY

Computer technology is now being used in modern fish-processing plants for continuous weight control and for administrative purposes. However, its application for quality assurance and control is still not established.

4.1 Grading

Grading and sorting of fish in every step of processing are the buzzwords in a modern industry concerned by quality assurance. This is not a simple task, as high speed is needed (1–10 fish/sec according to size), the environments are harsh and wet, and there is a lot of variation in size, species, and products. Most of the sorting is done manually. Grading after weight is done on a static scale. Automatic grading by use of dynamic scales, for measuring weight, and by the use of mechanical devices and/or light emission diodes, for measuring length and/or thickness, has been introduced.

Application of vision technique might have a future here because of its accuracy in measuring the length of the fish, the speed, and the possibility of differentiating between species. Vision techniques are also researched for trimming and portioning of fillets after weight in combination with a sensor-guided robot. If the fish is not length sorted on board (also by weight), sorting before processing on filleting machines is necessary in order to optimize the yield. As mentioned (1), this can be done by the use of vision techniques.

4.2 The Fish-Processing Plant

Let us look at the flow of fish at a processing plant (Figure 3) and point out the critical points for quality assurance, taking into consideration the possible methods for predictive, in-line control measurements.

A natural precondition is that the plant has a rational layout with appropriate chilling facilities, and that it is fully mechanized with weight control and the necessary hygienic control.

The Chill Store

The raw fish material can either be bought at an auction or landed directly at the processing plant. In Scandinavia it will most often be iced in boxes, but containerization in ice or in iced sea water (800–1200 L) is encountered. The fish might also be iced on shelves.

According to normal practice the fish is transported directly to the chill store in boxes from the auction or from the boat. The chill store is a very simple room, and the production planning is often "first in, first out," as the fish boxes are organized by date of catch (= age by arrival) if known. Spot tests to control the quality might be conducted. Transportation is via truck. In order to be able to use the information from the label on the fish box, secured

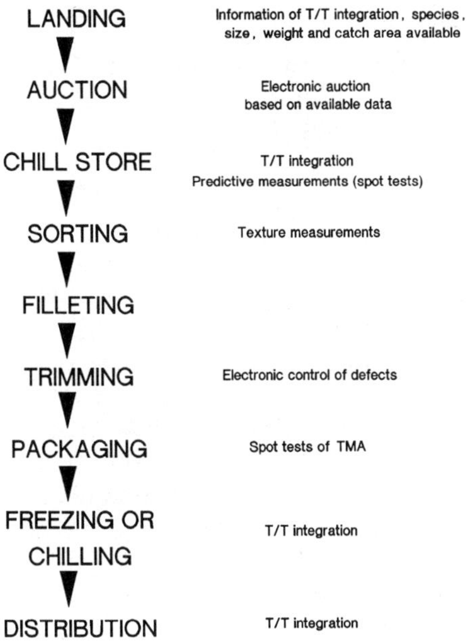

Figure 3 Flow sheet for a processing plant using integrated quality assurance. The left column shows the manufacturing flow. The right column shows the available information, possible use of this, and further steps for complete documentation of the quality for the final product.

by the quality assurance at sea, it might be necessary to reorganize the chill store for raw fish, or even to build an automated store and to store the information in a main computer to be used together with the sales forecast.

Time/temperature (T/T) integration is necessary in the store in order to know the combined strain on the raw material.

In the future plant it will also be possible to carry out measurements, which can predict the remaining shelf life based on microbiological counts of specific spoilage bacteria or on sensory analyses. Measurements of dry matter and fat might also be necessary for production planning.

Processing and Distribution

Even with the suggested control system in the chill store, misjudgements due to over- and/or undercapacity in the processing plant are possible, and an in-line grading after quality assurance would be useful before processing at the filleting machines.

It will presumably take many years before a physically robust "artificial nose" will be ready for use in fish grading. A more obvious solution could be grading with respect to texture, as fish of low quality often will have a soft texture.

After filleting and perhaps portioning, the products are packed or either chill-stored or frozen before distribution or further processing. In this last step at the plant, the time/temperature integration will again give valuable information. During all steps of these operations, the information should be registered and added to the information labels of the different fish batches.

Further control of the product is not necessary. However, control analyses may be required for trade purposes, including microbiological counts and total volatile bases (TVB) or trimethylamine (TMA).

5. TEST SYSTEMS FOR FISH QUALITY

A review of different methods for measuring a quality index of fish freshness is given by Jespersen (9).

Referring to the flow in a processing plant, there is a need for:

- Predictive methods
  - microbiological
  - sensory
  - chemical
  - physical
- In-line measurements
  - "quality"
  - texture
  - defects
- Control methods
  - quick

A test system should give a linear relationship with time of storage as well as results that bear the same relationship to temperature as the decline in freshness.

The requirements for a sensor/sensor system for this test system, to be used in-line in fish processing, are extensive:

- Real-time dynamic response
- Multiprocessing of information
- Noncontact
- Accurate and repetitive
- High mean time between failures (MTBF)
- Low price
- Nontoxic
- Sterilizable
- Insensitive to electrical/electromagnetic interferences
- Physically robust

The demand for noncontacting is a problem, and only acoustic, optical, and electromagnetic sensors can fulfill this requirement. In the fish industry the most plausible sensor is based on vision techniques.

The much-discussed biosensors for measurement of gases are very sensitive; however, development is still only at a laboratory level.

Most often in the fish industry it is necessary to utilize off-line measurements on spot-test samples in a laboratory. Typical examples of quick off-line methods are microwave, nuclear magnetic resonance (NMR), near infrared reflectance spectroscopy (NIR), high performance liquid chromatography (HPLC), gas chromatography (GC), and flow injection analysis (FIA), which can give responses within an hour. They work destructively, e.g., the sample is minced before the measuring. The typical use is for dry matter, fat, nucleotides, and TVB, TMA, or dimethylamine (DMA).

5.1 Predictive Methods

Predictive Microbiology

The predictive *microbiological* methods are not reviewed here, but it should be mentioned that it is possible to make a mathematical model predicting shelf life for chilled fish in ice based on the microbiological quality of the fish, estimated rapidly by conductance assays (13,14). Further work is needed, however, before a model can be made for vacuum-packed fish and lightly preserved products.

Predictive Sensory Methods

The Tasmanian Food Research Unit (TFRU) system, originally developed in Australia (2), is a very quick *sensory* method, which can be used for prediction of shelf life. The system is based on known principles of testing appearance, feel and smell, which gives an overall picture of the quality of the fish. The system is based on demerit points; the highest score is given for the poorest quality, as people generally are best at pointing out defects. Fish obtaining a low score is therefore of the best quality.

A species-specific number of attributes has to be judged, and the score ranges for each attribute vary from 0–1 to 0–3. Attributes of lesser importance can only be scored 0 or 1. Due to this differentiation, a single or some few misjudgments have little impact on the total test score. The system does not therefore lead to rejection of a sample based on one attribute only.

Each species of fish has its own characteristic quality parameters, and it is therefore necessary to use a species-specific evaluation system. This is the case even for closely related fish such as cod, whiting, and haddock. An example of the evaluation of cod is shown in Figure 4. The system is designed to give a linear relationship with time. The quality can therefore be expressed as the number of days in storage, under the best possible temperature conditions in ice at 0°C.

To predict remaining shelf life as days in ice for fish with unknown prehistory, this system can be used to evaluate the normalized time-temperature data for the sample (equivalent number of days in ice), where shelf life in ice is calculated in terms of days on ice to reach the maximum score.

The system has been modified in Denmark as part of a Ph.D. thesis project between the Technological Laboratory of the Danish Ministry of Fisheries and Matcon Consulting Engineers and Planners Ltd., and it is used for cod and herring. The system has been implemented on a Casio FX-829 pocket calculator and is now being incorporated into a portable hand-held terminal,

Quality parameter	Character		
Appearance	Appearance	0	V. bright
		1	Bright
		2	Sl. dull
		3	Dull
	Skin	0	Firm
		1	Soft
	Stiffness	0	Rigor
		1	Post-rigor
	Slime	0	Clear
		1	Unclear
		2	Sl.cloudy/milky
		3	V.cloudy
Eyes	Clarity	0	Clear
		1	Sl.cloudy
		2	Cloudy
	Shape	0	Normal
		1	Sl.sunken
		2	Sunken
Gills	Colour	0	Characteristic, red
		1	Sl.faded
		2	Faded, discoloured
	Smell	0	Fresh, seaweed/metallic
		1	Fishy
		2	Stale
		3	Spoilt
	Mucus	0	Absent
		1	Moderate
		2	Excessive
Flesh colour	In open surfaces	0	Translucent
		1	Grey
		2	Yellow-brown
Blood	In throat	0	Red
		1	Dark red
		2	Brown
Character sum			

Figure 4 Revised TFRU method for quality assessment of cod.

which can be connected to a PC for data collection. The terminal asks the assessor the relevant questions and guides him through the entire assessment. When the evaluation is completed, the sum of score points is displayed on a small screen. This information might be converted to EEC standards. It is also possible to incorporate relevant information such as catching area, date of catch, numbers of boxes, etc., and all this information can either be transferred to a main computer or to a printer.

The TFRU system can be used in the chill store for production planning but could also be a valuable tool for the fish inspector or even for the fisherman.

Up to now there has been no successful research on predictive *chemical methods*, but the above-mentioned combination of time/temperature integration and condition factor can give a *physical* prediction of shelf life and yield.

Use of Vision Techniques in Future Predictive Measurements

In the described sensory methods for testing of raw material, most of the parameters are based on a visual impression. This might be the starting point for an objective method using vision techniques. Changes in color and presence of slime on the skin, color and slime of the gills, and the changing transparency of the eyes are an index of fish freshness. The changes can easily be seen and might, with an appropriate illumination, be recorded by a color video camera. However, a lot of research is needed to produce a specific system, which can be used for in-line control of fish.

5.2 In-Line Methods

Research is needed on noncontact in-line measurements which can work very fast (1–10 fish/sec) in relation to an inspection of each individual fish.

Quality

Vision techniques might, in the future, provide us with a measurement of what we can see when fish deteriorates. A promising technique might be measurement of fluorescence. A combination of scattered data in the literature and recent work at the Technological Laboratory of the Danish Ministry of Fisheries is showing a relation between fluorescence of fish at 466 nm, when excited by light at 358 nm, and freshness.

Texture

The texture of the fish is often a good measure of the quality and might be determined in several ways, most often in shear or compression cells. Future developments, however, might give contacting mechanical or acoustic methods to be used in-line.

Defect	Source of light
bone	UV light (Huss et.al., 1985)
parasites	UV light
blood	visible light with color filters (Emsholm et.al., 1987)
gaping	visible light with color filters (Emsholm et.al., 1987)

Figure 5 Methods for detection of fillet defects.

Defects

Today fillet defects can be detected automatically, using video camera techniques and subsequent image analysis. The presence of bone fragments can be detected using UV light as the bones exhibit an autofluorescence. A system for fish bone detection based on fluorescence is commercially available, and other systems are being researched.

As shown in Figure 5, other kinds of defects such as blood spots and gaping can also be detected using vision techniques and image analysis. Investigations of Emsholm et al. (5) indicate that it is feasible to develop a mechanical device for automatic in-line sorting of fillets with use of visible light and color filters.

5.3 Control Methods

Recent developments in the food industry provide quick methods for measuring water, lipid, protein, TVB and the standard analytical measurements. It is characteristic for those quick methods that expensive equipment is needed, such as for NMR, NIR, HPLC, and GC.

At the Technological Laboratory we have developed quick methods using flow injection techniques on TVB, TMA, and DMA (6). A quick distillation method (15 min/sample) for TVB determination has recently been developed. Using the flow injection analysis, however, 120 samples an hour can be determined.

Nonsegmented continuous flow techniques were introduced 13 years ago by Ruzicka and co-workers (15) and were called flow injection analysis (FIA). FIA involves the injection of a liquid of a sample into a moving nonsegmented continuous carrier stream of a suitable liquid. In the determination of TVB, TMA, and DMA, gas diffusion of the analyte from the donor stream across a permeable teflon membrane into a recipient stream is used, where the product of reaction is detected by a spectrophotometrical device.

Wekell et al. (16) has also reported the use of FIA for TVB determination.

The FIA method is correlated with the picrate method (3) and GC (10) for DMA and TMA, the Conway method for TMA and TVB.

It was concluded that FIA is applicable for TVB, TMA, and DMA determinations, and that further investigations should be done in order to make the method commercial.

6. SUMMARY

A precondition for a quality assurance system in the fishing industry is optimal on-board handling of the chilled fish. It is also necessary that fish in every link in the chain between catch and consumer be handled as efficiently as possible.

A fair price is one way to guarantee quality. We are, however, still missing commercially available equipment to measure the length with the right precision. Measurement of the length is necessary to calculate the condition factor, which might be used to predict yield and thereby price. A system based on vision techniques might be the solution.

Other promising areas where vision techniques might be used in the future are detection of fillet defects, grading of species, measuring of freshness index, and measuring of weight for portioning of fillets.

REFERENCES

1. Arnarson, H., K. Bengoetxea, and L. F. Pau. 1988. International Journal of Pattern Recognition and Artificial Intelligence 2(4): 657–671.

2. Branch, A. C., and A. M. A. Vail. 1985. Food Technology in Australia 37: 352–355.
3. Castell, C. H., B. Smith, and W. J. Dyer. 1945. J. Fish. Res. Bd. Can. 31(I): 383–389.
4. Conway, E. J., and A. Byrne. 1933. Biochem. Journ. 27: 419–429.
5. Emsholm, H., J. Nielsen, and T. Børresen. 1987. *Proceedings of the 4th European Conference on Food Chemistry*, Loen, Norway, 1–4 June 1987, Ås, Norwegian Food Research Institute, pp. 450–459.
6. Garn, M. B., and C. Nystrup. 1985. M.Sc. thesis (in Danish), Danish Ministry of Fisheries, Technical University, Lyngby, Denmark.
7. Heldbo, J., H. Arnason, and G. L. Smith. 1990. *Proceedings of the 34th Annual Atlantic Fisheries Technological Conference*, August 27–30, St. John's, NF, Canada.
8. Huus, H. H., P. Sigsgaard, and S. A. Jensen. 1985. J. Food Protection 48(5): 393–396.
9. Jesperson, C. M. 1988. *Kvalitet og fiskeforædling. Kvalitetsbestemmelse af fisk og fiskeprodukter.* Danish Ministry of Fisheries, Technical University, Lyngby, Denmark.
10. Lundstrøm, R. C., and L. D. Racicit. 1983. J. Assoc. off. Anal. Chem., 66(5): 1158–1163.
11. Maack, P. K. 1985. *Proceedings of a Jubilee Seminar*, Sept. 23–24, 1984, A. Möller (ed.), Icelandic Fisheries Laboratories, Reykjavik, Iceland.
12. Mortensen, B. 1988. M.Sc. thesis (in Danish), Danish Ministry of Fisheries, Technical University, Lyngby, Denmark.
13. Ravn Jørgensen, B., D. Gibson, and H. H. Huss, 1987. Paper presented at the 17th Annual WEFTA Meeting in Dublin, 1987.
14. Ravn Jørgensen, B., D. M. Gibson, and H. H. Huss. 1988. Int. Jour. Food Microbiol. 4: 295–307.

15. Ruzicka, J., and E. H. Hansen. 1984. Analytical Chimia Acta 161: 1–25.

16. Wekell, M. M., T. A. Hollingworth, and J. J. Sullivan. 1986. In *Proceedings of an International Symposium on Seafood Quality Determination*, in Alaska, November, Elsevier, Amsterdam.

17. Kramer, D. E., and J. Liston (eds.). 1987. Devel. Food Sciences 15: 17–26.

2

Vision Applications in the Fishing and Fish Product Industries

HÖRDUR ARNARSON *Marel HF, Reykjavik, Iceland*

KEPA BENGOETXEA *Dow Chemical Company, Tarragona, Spain*

L. F. PAU *Technical University of Denmark, Lyngby, Denmark*

◆◆◆

1. Introduction
2. Classes of Applications
    2.1 Fish Sorting
    2.2 Fish Product Sorting
    2.3 Internal Defects
3. Optical Properties of Fish Classes
    3.1 Skin
    3.2 Flesh
4. Sensors and Environment
    4.1 Environment
    4.2 Sensors
5. Fish Sorting
    5.1 Needed Performance

- 5.2 Illumination
- 5.3 Algorithms
6. Fish Product Sorting
   - 6.1 Needed Performances
   - 6.2 Illumination
   - 6.3 Algorithms
7. Internal Defects
   - 7.1 Needed Performance
   - 7.2 Illumination
   - 7.3 Algorithms
8. Surface Defects
   - 8.1 Needed Performances
   - 8.2 Illumination
   - 8.3 Algorithms
9. Specific Hardware and Architecture Requirements

References

1. INTRODUCTION

The fishing and fish product industries worldwide are facing tremendous challenges from more stringent health testing requirements as well as overriding manpower cost and unavailability increases. This is related to the fact that these industries still use manual manpower almost exclusively for all basic operations, and, at the same time, consumers simply do not tolerate any longer product imperfections (bones, worms, blood spots, irregular shapes). Automatic factories are mandatory because manpower cannot be found, e.g., at candling tables.

At the processing level, the machines (filleting, head cutters), while operative, still cannot handle in the same batch diverse fish types and/or sizes. Any setup due to such variance is both time consuming and costly.

It is only very recently that, on one hand, researchers have discovered the potential use of machine vision in the fish industry and, on the other hand, that vision system performances and prices are approaching ranges where they are acceptable to the industry. By and large, the application of vision in the fishing and fish product industries is a virgin area with very little being investigated so far with the exception of sorting (size or color) by diode arrays (e.g., shrimp sorting, or position sensing in filleting machines) or surface bone detection.

It is the purpose of this chapter to present the findings of a significant research program carried out in collaboration between university and industry. Section 2 will categorize the classes of problems where vision is applicable. Section 3 in turn classifies the fish species in terms of their optical properties, which the vision system sensors and illumination systems will be built around. These optical properties, as well as the operational environments (fishing trawlers, fish product factories, aquaculture dams) lead to constraints on sensor types and characteristics, as discussed in Section 4. Sections 5 to 8 will, for each application class, highlight needed vision system performance, illumination, image processing algorithms, and results. Specific vision hardware and architecture requirements are summarized in Section 9.

It should be emphasized that, whereas algorithms to be discussed are rather straightforward, all other implementation aspects are highly delicate for objects as diverse as fish in the harsh environments considered.

2. CLASSES OF APPLICATIONS

Whereas the application of remote sensing to spot fishing zones and photocell-based sorting are outside the scope of this chapter, the main classes of application of vision to the fishing and fish product industries are the following.

2.1 Fish Sorting

This operation aims at grouping fish by species, length, and/or weight for further processing.

Sorting fish in water is required in fish farms (in dams, or in sea enclosures) to count fish, to estimate their weight from their length, and to estimate the growth curve for a population/batch of fish. In this application, the individual fish cannot be tagged, as sampling takes normally place in terms of the imaging process.

Sorting dead or dying fish in air is required aboard fishing boats for packaging purposes. Typically, the fish or shrimp caught in a net must be sorted by species, length, and/or weight, and special cases set aside, before they go into boxes compatible with the auctioning process (from the boat or on landing). If fish processing or freezing takes place on board, the sorting requirement is still maintained. The sorting can be combined with weighing by special purpose digital weights often imbedded in transport systems. In this instance, each fish can be tagged, with attributes such as destination, weight, type, length, thickness, color, shape. These attributes can in turn be used as inputs to the package-handling systems and/or to the fish-processing machines. One special subprocess is to estimate the yield of the fish-processing machines (in terms of wasted/rejected fish meat) based on the adequacy of shape and length parameters and the machine setups. Similarly, it is necessary to optimize the packaging density.

2.2 Fish Product Sorting

The fish-processing machines produce, even within a given batch, highly irregular products (typically fillets, but also water jet cut fish cubes or fish slices). This is essentially due to the high-speed mechanical handling of nonsolid fish meat, resulting in irregular shapes, lengths, thicknesses, as well as in fillets folded together, overlapping, etc.

The fish product operation consists of dispatching the fish parts according to shape and weight, and also carrying visual quality control of these parts for shape and size compliance. Odd fillets (elongated and too long, or too small) must be eliminated; connected fish halves must be aligned for further processing; bent fillets must be flattened out; fish cubes must be eliminated if they do not comply with the shape requirements; fillets must be separated if they are to be weighed individually.

2.3 Internal Defects

This class of applications consists of carrying out quality control, especially with respect to health hazards due to the fish meat itself.

- *Worm detection*: Cod, whitefish, salmonids, and other species can carry various types of larvae such as *Trichinella spiralis*. Some prior work has gone into designing optical and other detection systems to supplement or replace backlighting of the fillets (which is today a tiring human inspection process) (see Refs. 18–21). If, however, the fish were frozen before, the worms would be dead.

- *Detection of bone residues*: Whereas fish product machines should normally eliminate most bones, they do not always, thus raising the issue of acceptance level for bone residues among consumers. Special detection processes have been implemented using either the self fluorescence of bone residues or soft X-ray detection.
- *Fish liver*: Locating the liver inside the fish for extraction, and possibly finding parasites on the extracted liver (2) is a specific problem. The skin on the liver nourishes the parasites.
- *Fish meat texture*: It is necessary for some types of fish products, to estimate as early as possible the chemical and texture properties of the fish meat. This applies to the fish mince products. One approach is rheological imaging with fish tissue pH correlation.
- *Surface defects*: Consumer acceptance is related to the aesthetics of fish or fish products, even if no health hazard is involved. Here again the fish-processing machinery, especially washing, should take care of it, but defects are nevertheless frequent. Blood spots on whitefish fillets as well as on tuna slices, require color discrimination (black tuna meat results in bad taste in canned or smoked products). Acceptable colors are well known and subject to many effects such as water dilution, blood aging, freezing, etc. Codex standards exist for acceptable surface defects in terms of the absolute area of such defects (typically 3, 5, 10 cm^2 intervals).
- *Miscellaneous*: Other applications of vision to fish products include: water quality monitoring by particle diffuse scattering, for on-line control of alevin farming units, and thermal control of water in fish farms to regulate the water renewal rate; fish dynamics to track bank movements through sonar systems or hydrophone arrays.

3. OPTICAL PROPERTIES OF FISH CLASSES

There are about 20,000 living species of fish in the world, but we will consider in this paper only those of economic importance. They are found in the group of bony fishes (this group includes by far the majority of the living fishes) such as the herring-like fish, cod group fish, mackerel family fish, and flat fish.

The purpose of this section is to review the optical properties of skin and flesh of different species in order to build the illumination part of a vision system (see Table 1 in Chapter 4).

3.1 Skin

The outer skin or epidermis of a fish secretes a large amount of mucus which serves partly to reduce the frictional resistance of the water and partly as a barrier to parasites (1). Furthermore, the body of a bony fish, is covered more or less completely by scales (1). The distribution and possible existence of the scales should be considered, because the scales can give the same effect as a mirror. In several species of the tuna group, only the breast region has an armor of large scales, while the rest of the body and the tail are naked. In other species, the skin is furnished with large bony thorns or is armored with bony shields. Scales are lacking on very young fish, and are only developed when the fish has reached a certain length. After that the scales grow with the fish and they form rings. Because of these two characteristics—mucus and scales—fish should be considered as a shiny object for vision applications.

3.2 Flesh

Fish flesh has a relatively poor supply of blood and is usually white. However, fast-swimming fishes have a more rapid metabolism and a better blood supply to their muscles (1). Their flesh therefore has dark areas or is reddish (e.g., in the tuna).

4. SENSORS AND ENVIRONMENT

4.1 Environment

The environment in the fishing factory is very harsh, and on board fishing ships it is even worse. The main factors are illustrated in Table 1.

Because of this environment, the casing equipment is usually made out of stainless steel, with an accuracy demand of 0.05 mm, and must be strong enough to withstand washing.

Table 1 Fish Inspection Environment

Temperature	
	The temperature is usually between 0 and 20°C and in some places this variation can give the sensor problem because of its varying characteristics with temperature. A temperature between 15 and 20°C is unwanted because it decreases the quality of the fish; automation would allow one to lower the temperature.
Humidity	
	High humidity (90–95%) causes problems for electrical equipment.
Water	
	In fishing plants and on board fishing ships, almost everything is wet, so watertight equipment is needed. The reflection of the water also causes illumination problems.
Biological acids	
	These acids destroy things like rubber and some metals. Also the transport belt can change colors because of these acids.
Big motors	
	Big electrical motors are running in fishing plants and on board ships. These motors cause electromagnetic fields, which may influence cameras (video cameras).
Washing	
	Performed for health reasons and usually with highly pressurized waterjets.
Space limitation	
	On board ships there are severe space limitations. Vision equipment has to fit into the production lines, or these have to be reorganized as is very often necessary.

4.2 Sensors

Several types of sensors can be used in the fishing industry (Table 2), but they are highly dependent on each application. The most common sensors are solid state sensor (CCD and photodiode arrays), video cameras (vidicon), and line scan cameras (30–32). When selecting among them, several characteristics have to be taken into account.

- *Shape of sensor*: Although line scan cameras are well suited for industrial use (27), they cannot be used on board ships, because of the irregular movement of the ship, and they require synchronization between the camera and the transport belt.
- *Readout speed*: This is specially important in fish sorting, where high speeds are often needed. Video cameras offer higher readout speed than solid state cameras.
- *Dynamic range*: This is important in fish inspection, where a high gray scale resolution is needed. Solid state cameras generally have higher dynamic range than video cameras.
- *Spectral sensitivity*: This is especially important in fish inspection where only selected wavelengths of light are used.

Table 2 Needed Performance of Sensors

	Fish sorting (± 1 cm)	Fillet sorting (± 1 cm)	Worms, bones	Surface defects
Resolution (pixels)	700 × 500	600 × 400	700 × 700	500 × 500
Dynamic range (rms noise)	1000:1	1000:1	10000:1	5000:1
Frame rate (frames/sec)	300	300	25	25
Spectral sensitivity (nm)	390–770	390–770	UV, X-ray infrared	390–770

Table 3 Speed in Fish Sorting

	Image sensors	Nonimaging sensors (photodiodes)
Sorting by length and/or weight	12000 fish/hr	23000 fish/hr
Sorting by species	7000 fish/hr	Not possible
Sorting by shape for known species	2000 fish/hr	6000 fish/hr

- *Camera sync*: It has to be possible to control the camera from the computer because of needed synchronization with lighting, other cameras, and the movement of the transport belt.
- *Robustness*: Here solid state cameras offer clear advantages over video cameras.

Other important characteristics are sensitivity, lag, sensor distortion, operating voltage, size, weight, and price.

Because of their robustness and high dynamic range, solid state cameras are often chosen for use in the fishing industry. But in applications where high readout speed or high resolution are needed video cameras still offer advantages over solid state cameras (30,31).

5. FISH SORTING

This section covers the applications described in Section 2.1.

5.1 Needed Performance

For sorting live fish in water, the required speed is approximately 2000 fish/hr/sensor (Table 3). This involves contrast backlighting through water and fish separation by gates and water flow control in the inspection tunnel. For sorting fish in air, the speed is dictated by the supply rates (highly irregular) and by the rate at which sorting data must be used. An estimate is: This assumes a conveyor speed of 10 m/s, 4 bucket lines per set, 6 classes of lengths (or 6 packaging destinations), and 3 grades of shape compliance by species.

5.2 Illumination

For sorting live fish in water, backlighting is the best method, but two points should be considered in order to find a compromise between: 1) fish dynamics; fish is attracted by light (1), so it tries to stay in the place where the light is if the water flow is not strong enough; 2) turbulence of water; if the water flow is not laminar, it might interfere with the lighting.

The purest known natural water, Crater Lake in Oregon, is most transparent at 420 nm and absorbs strongly in both the near-ultraviolet and at wavelengths longer than 500 nm (2), but in fresh water there is an extremely wide range of water colors which depend mostly upon the ecology of the surrounding terrain (3) (e.g., blue, blue-green, or red-brown). When the water warms up in the spring, there is usually a great increase in chlorophyll-containing phytoplankton, which tends to shift the spectral transmission peak of the water toward that of chlorophyll at about 540 nm. In the winter, with little biological activity, the water clarity increases with a resulting shift in transmission toward blue.

The dissolved salts in seawater have little or no effect on the spectral absorbance of the water (4,5). However, other dissolved and suspended substances, notably chlorophyll and the yellow products of vegetable decay (6), do exert a profound influence shifting the spectral transmission of the water to about 530 and even beyond 570 nm.

For sorting fish in air, either diffuse direct light or backlighting could be used. According to Section 3, backlighting gives fewer problems except for the megrim, because of its thin body. It would be interesting to study which wavelength gives the maximum transmission through the translucent layer on the conveyor belt. Direct light requires a special setup to reduce glinting and avoid shadows even on specimens with no scales because of the silvery belly of the fish (Fig. 1).

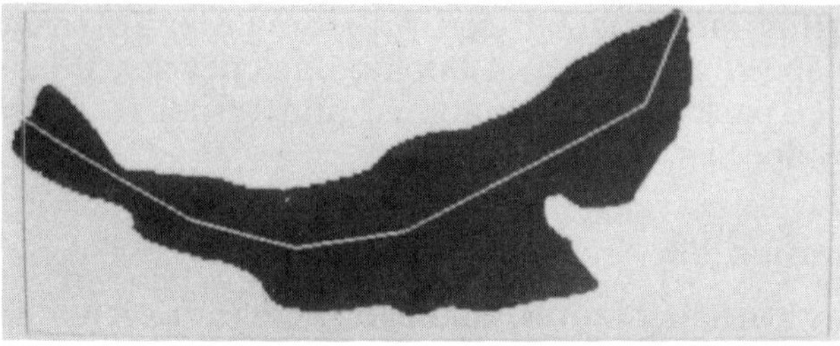

Figure 1 Threshold fish image taken on transport belt, with length estimation line.

5.3 Algorithms

Classical high-speed sorting procedures apply for moving parts. They involve picture or line scan binarization and exploiting the elongated shape of all fish to generate a fish characterization by a variable number of 2D or 3D slices or to known intervals in space and time. These slices are obtained by sensor scan control in connection with the transport, irrespective of fish orientation. Sorting by species as well as precise length estimation involves hypothesizing head and tail locations each time an object is detected and carrying out morphological filtering of selected windows at each extremity (Fig. 2). The filter kernels, as well as feature combination rules, must be size independent but species dependent. The shape anomalies between male and female fish offer interesting challenges, as do unretracted fins.

6. FISH PRODUCT SORTING

This section covers the applications described in Section 2.2.

6.1 Needed Performances

Fish product sorting speed is dictated by the output rate from filleting, chopping, waterjet cutting or other machinery (including inspection systems of type 2.3 or 2.4). Such speeds are thus, per

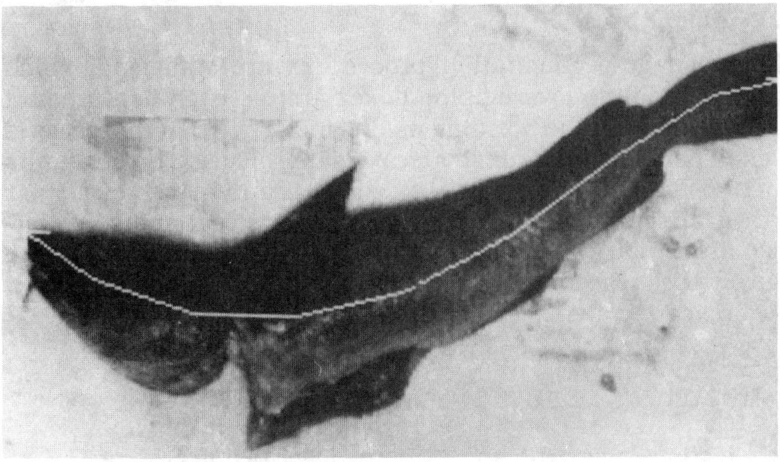

Figure 2 Length estimation of fish (note compensation for fins).

line, at most 10000 products/hr for medium-sized fish: the size measurement errors per item must be ± 1–3 cm in any direction, and the thickness measurement error ± 0.6 cm. The speed of the products on the transport can be 0.8–3 m/sec. One special case is head cutting, at 2500 fish/hr, with the alignment required by this task. Another is the longitudinal alignment of herring at 4000 fish/hr with a 5° angular tolerance.

6.2 Illumination

For estimation of length and shape, either diffuse light or backlighting can be used. Direct light requires a change of background in the case of white fish species or reddish flesh species, so it appears that backlighting is best for all species. On the other hand, backlighting needs a translucent layer on the conveyor belt which can diffuse the light, thus producing shadows which can make difficult the determination of shape and length. Current recommendations by the Fisheries and Oceans Administration in Canada involve a 5-mm thick acrylic sheet with 45% translucency (7).

6.3 Algorithms

Simple image segmentation procedures are sufficient, which draw minimal nonoverlapping boxes around each fish product and characterize the box dimensions in 2D or 3D (if thickness is needed using an additional lateral vision sensor). The box limits can be estimated by very high-speed line scanning perpendicular to the transport direction. The acceptance rules than use knowledge about sorting criteria, and about the upfront machine, to classify the tagged boxes into a few classes; counting and weight labeling can be performed simultaneously.

7. INTERNAL DEFECTS

This section covers the applications described in Section 2.3.

7.1 Needed Performances

The rates at which internal defects are manually detected, correctly or not, are today so bad (when the need for detection is present) that any improvement is very much looked forward to. At the same time, worse detection performances will not be accepted, neither will variability due to, e.g., fish thickness, fish age, freezing process. Especially cumbersome is the limitation of vision systems to limited depth penetration and to inspection one side at a time only (thus raising the unsolved need for a fish face-turning system). A very rough estimate of nondetection rates by manual/visual methods is 20% for worms, 10% for bone residues. The speed requirement is in general identical to the one in Section 6.1, with a median around 5000 products/hr for residue bones.

7.2 Illumination

Bones

Considering the great penetration power of the X-ray (8) and the fact that X-rays are highly attenuated by bones (9) the collimated X-ray backillumination is nowadays the best lighting for the detection of bones; even relatively small bones can be detected. For various reasons, X-ray energy should be less than 6 KeV (10).

Huss et al. (11) showed the autofluorescence phenomenon of fish bones. All fish species examined exhibited a characteristic autofluorescence at 390 nm when excited at 340 nm, but this phenomenon can only be used if the bones are at the surface but not inside the flesh. The relative fluorescence intensity decreases about 50% when the bones are covered with 1.0 mm of flesh.

Worms

For the detection of worms, especially those inside the flesh, there is presently no special lighting which can be used with enough robustness of detection. Alternatives are candling tables, laser, ultrasound, neutron radiography, and X-ray fluorescence.

Candling tables are limited because of the thickness of the fillet. Power (12) found that for efficient candling, an incident light value of 3–5 foot-candles and a transmitted light value of approximately 50 foot-candles should be used. Under the best conditions, he was able to find only 73–74% of the worms in heavily infected cod fillets. He also observed, that worms embedded deeper than 6 mm below the surface, or in flesh with discoloration, or under skin (ventral skin or sarcolemma), are unlikely to be detected by candling.

Experiments with ultrasound shows that it might be used for the detection of worms, but the equipment is too expensive and the technique is insufficiently known. Hafteinsson (13) found that sealworms can be detected in cod fillets as thick as 4.5 cm with skin. They used a new SLAM, the Sonomicroscope-System 140, which can be operated at 10, 24, 30, and 100 Mhz frequency. The collagen content of the sealworm is probably responsible for the fact that the detection is possible in such thick fillets. But the collagen content of the skin is very high too, which makes difficult the detection of sealworms in fillets with ventral skin or sarcolemma.

The high sensitivity of thermal neutrons (0.01–0.03 eV) to hydrogen content variations indicates that neutron radiography might be useful in some medical and biological problems (9). This property might be applied to the detection of worms. Anyway, the instrumentation would be very expensive, and serious research should be conducted in this direction. Besides, neutrons are easily transmitted through bone, while X-rays in the medical energy region are highly attenuated by bone. Therefore, neutron tech-

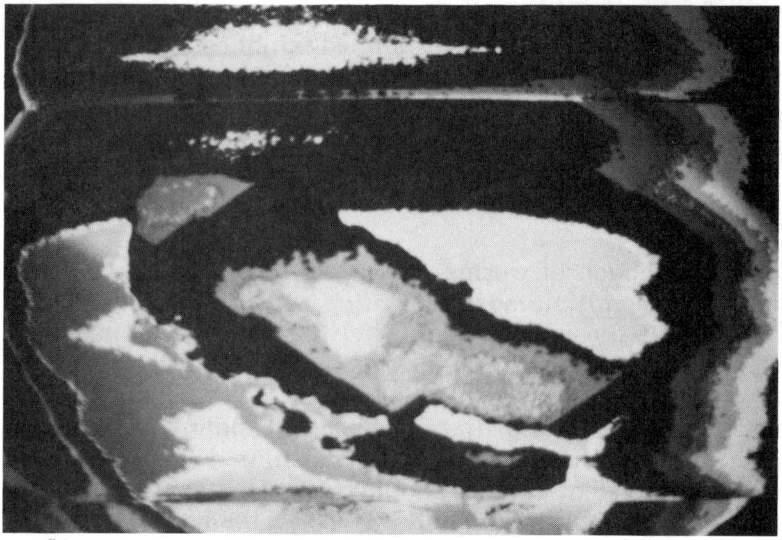

Figure 3 Worms in fillets by laser imagery (visible, laser images). *Source*: L. F. Pau and Battelle Institute (unpublished, 1986).

niques offer the possibility for viewing tissue which might be masked by bone in a X-ray examination (14) by which means worms under bones could be detected.

Fluorescence of X-rays around 100 KeV might be suitable for the detection of worms because of the great penetrating power of X-rays. Anyway, such a system has not been studied so far. Figure 3 gives promising results for laser transmission scanning.

7.3 Algorithms

As the key is the optical defect detection process, only image enhancement through look-up tables and video contrast stretching are used prior to binarization.

8. SURFACE DEFECTS

This section covers the applications described in Section 2.4.

8.1 Needed Performances

Statistical lot acceptance criteria apply to surface defects, but it must be said that even trained operators are not consistent in the delineation of the defect boundaries or color features. Moreover, here again two-face inspection is required. Human correct detection rates are high, but result usually in high false alarm rates, which lead to discarding acceptable products. Other surface defects are fat, as well as fish sex in the case of salmon and "loddes". Also, black skin and blood spots may overlay. The nondetection and false alarm rates on products with defects should be less than 5%. The speed requirement is identical to the one in Section 6.1, with a median around 5000 products/hr.

8.2 Illumination

The best lighting for the detection of blood spots and black skin is by diffuse direct light. Special care must be taken to reduce shadows and glinting, and to enhance the discrimination between blood spots and black skin in relation to the flesh, and likewise flesh in relation to the background. For this purpose, it is very important to carry out a study of which wavelength bands should be enhanced considering the emission spectrum of the lamp as well as color filter bands. It was found that a discharge lamp (with maximum components in the green part of the spectrum in a diffuse environment) and a green filter (to cut components in the yellow and red parts of the spectrum) made up the most suitable setup for the detection of blood spots and black skin (15).

8.3 Algorithms

This involves region area estimation, after recursive boundary estimation to remove uncertain boundary pixels. Simple region erosion operators (by blurring and shrinking, or by morphological filtering) are sufficient (Fig. 4) (Table 4).

9. SPECIFIC HARDWARE AND ARCHITECTURE REQUIREMENTS

The application of vision to fishing and fish products industry involves a number of specific requirements often not met by cur-

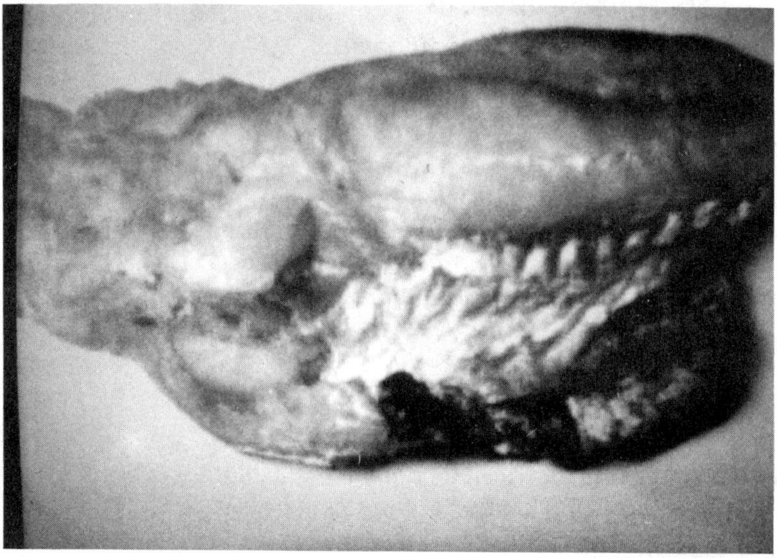

Figure 4 Blood spots (middle) and black skin (lower right side). Visible light (right) and filtered image (left).

rent commercial systems, besides those sensor and illumination requirements discussed in Section 4.

Such requirements include:

- Concurrent processing of two sensor channels with digital ratios, which involves changes in current color vision systems
- Custom gate arrays for shape classification and precise length estimation
- Asynchronous sensor (video or laser) scan control and control of parallel spaced scans

Table 4 Illumination of Defects

	Electromagnetic radiation								
	Visible light and laser		IR light	UV light		X-Ray		Neutron	Ultrasounds
	Trans-mission[a]	Scat-tering	Trans-mission[a]	Scat-tering	Fluores-cence	Trans-mission	Fluores-cence	Trans-mission	Trans-mission
Black skin	Yes	Yes	Yes	Yes	No	No	No	No	No
Sarcolemma[b]	Yes	No	Yes	No	No	No	No	No	No
Parasites inside[c]	No	No	No	No	No	No	?	?	?
Parasites surface[d]	No	Yes	No	Yes	Yes	No	?	?	?
Bones inside[e]	Yes	No	Yes	No	No	Yes	?	No	?
Bones surface[f]	Yes	Badly	Yes	Yes	Yes	Yes	?	No	?

[a]Limitation: Thickness
[b]White skin covering the muscle is considered as a good part.
[c]Anisakis simplex.
[d]*Phocanema decipiens.*
[e]Collarbones.
[f]Pinboxes.

REFERENCES

1. Muus, B. J., and P. Dahlstrom. 1974. *Collins Guide to the Sea Fishes of Britain and North-Western Europe*, Collins, London.
2. Smith, R. C., and J. E. Tyler. 1967. J. Opt. Soc. Amer. 57: 589–595.
3. Lythgoe, J. N., and D. P. M. Northmore. 1973. *Colours Underwater*, Colour 73, Adam Hilger, London.
4. Clarke, G. L., and H. R. James. 1939. J. Opt. Soc. Amer. 29:43–55.
5. Sullivan, S. A. 1963. J. Opt. Soc. Am. 53:962–967.
6. Kalle, K. 1966. Oceanog. Marine Biol. Ann. Rev. 4:91–104.
7. Grading Standard Fresh and Frozen Atlantic Groundfish Products, DFO-1983-1, Inspect Tech. Branch, Ottawa, Canada, 1983.
8. Liebhafsky, H. A., H. G. Pfeiffer, E. H. Winslow, and P. D. Zemany. 1972. *X-rays, Electrons and Analytical Chemistry*, John Wiley and Sons, New York.
9. Brown, M., and P. B. Parks. 1969. Am. J. Roentg. 106:472–485.
10. Dyck, R. H. 1987. *X-Ray Imaging with Fairchild CCD Image Sensors*. Fairchild Weston CD Imaging Division.
11. Huss, H. H., P. Sigsgaard, S. A. Jensen. 1984. J. of Food Protection 48(5):393–396.
12. Power, H. H. 1958. J. Fish Res. Bd. Canada 15:537–542.
13. Hafsteinsson, H., and S. S. H. Rizvi. 1987. J. of Food Protection, 50(1):70–84.
14. Berger, H. 1970. *Neutron Radiography, Research Testing in Nondestructive Testing*, Academic Press, London and New York, pp. 269–314.
15. Bengoetxea, K. 1988. Report No. 497, Technical University of Denmark, Lyngby.

16. Quarmby, A. R., et al. 1982. Australian Fisheries J. 40(11):42–45.

17. Bulow, F. J., and R. D. Anderson. 1979. Prog. Fish. Cult. 41(2):73.

18. Odense, P. H. 1978. Fisheries and Marine Service, Halifax, Nova Scotia, Report No 106.

19. Wommark, M. R. Worm Detection Process, U.S. Patent 04061788 (1977).

20. Orlewski, R. 1975. Bull. Acad. Polonaise des Sciences, Sciences Biologiques 23(9):629–630.

21. Wetzel, R. FRG Patent 01673195 (1970).

22. Ddvison, A. 1979. North Atlantic Sea Food, Mac Millan Ltd. London.

23. King, J. E. 1964. Seals of the world, Trustees of the British Museum of Natural History, London.

24. A primer to parasites, Seafod Leader J:44–46, Winter 1986.

25. Davis, H. H., and P. Reece. 1976. J. Science Food Agriculture 33:1138–1142.

26. Rippy, J. H. C. 1970. J. Fishing Research Canada 27:963–965.

27. Batchelor, B. G., D. A. Hill, and D. C. Hodgson. 1985. *Automated Visual Inspection*, IFS Publication.

28. Gonzalez, R. C., and P. Wintz. 1977. *Digital Image Processing*. Addison Wesley Publ., Reading, MA.

29. Pau, L. F., M. Mahmoud, and El Nahas. 1985. *An Introduction to Infrared Image Acquisition and Classification*. Wiley/RSP, London.

30. Spiell, E. 1987. Laser Focus/Electro-Optics, January.

31. Loebl, J. 1986. *Choosing a Video Input*. Sensor Review, April.

32. Ballard, D. H., and C. M. Brown. 1982. *Computer Vision*, Prentice-Hall, Englewood Cliffs, NJ.

II

Biological and Optical Properties of Fish for Vision Applications

3

Optical Spectra of Fish Flesh and Quality Defects in Fish

JÓN PÉTURSSON *Science Institute, University of Iceland, Reykjavik, Iceland*

1. Introduction
   1.1 The Water Window
   1.2 Previous Work
2. Concepts and Experimental Methods
3. Results
   3.1 Optical Absorption
   3.2 Scattering
   3.3 Diffuse Reflectance
4. Discussion
   4.1 Fluorescence
   4.2 Effects of Geometry
   4.3 Laser Scanning
   4.4 Optical Properties of Tissue from Other Fish Species

   Appendix

   References

1. INTRODUCTION

Automatic sorting of fillets and other fish products according to the number of quality defects they contain is the next logical step toward fully automatic fish processing. This will require reliable and practical methods for detection of the various kinds of defects in fish flesh. Bones, cod worms, blood stains, pieces of skin, etc., are defects that can and must be removed before packing. Today the labor-intensive operation of visual detection and manual removal of these defects is a heavy burden on the fish-processing industry.

Other factors that also affect quality are the time from catch to processing and conditions during that time. These can result in downgrading of whole fish or even batches of fish. Gaping is an important example of this kind of defect. In this case it is not possible to improve the product by localizing and removing defects. The whole fish has to be sorted into a lower grade. Ideally this deterioration should be prevented by good treatment. In practice, however, it will always be necessary to monitor these factors in order to guarantee top-quality products.

At present, quality defects in fish products are mostly detected by human vision, i.e., optically. Automatic detection will probably also use mostly optical techniques, although acoustics or X-rays may possibly be used in combination with optical methods.

A lead in this field gives the fish-processing industry a stronger position in the international competition that is more and more centered on quality. Furthermore, development and production of machinery for this purpose will be a high technology in its own right.

This author started to investigate the cod worm problem in 1981, on the urging of Dr. Björn Dagbjartsson, who was then director of the Icelandic Fisheries Laboratories.

This work started with a study of available literature and of possible new detection methods. Some simple experiments with X-rays, both conventional X-rays and CT tomography, were carried out, and also with fiber optics. Detailed measurements of the optical properties of cod flesh, nematodes, and some other defects were undertaken. The research was carried out at the Science Institute in Reykjavik and financed by The Icelandic Codworm Committee and the National Research Council in Iceland.

Several references exist on research into candling methods. H. E. Power (1) investigated in detail the effects of different light intensities and various color filters on the effectiveness of candling.

In 1970 Dadi Àgùstsson (2) compared several kinds of lamps and studied ways to optimize candling illumination as well as the ambient light. His recommendations have been the basis for the design of candling tables to this day.

It may well be possible to improve on the design of candling tables in order to reduce eye fatigue. It is also important to check frequently the condition of candling table surfaces and light bulbs. An investigation in 1987 (3) into some recent types of fluorescent lights led to a recommendation to use type OSRAM No. 12 (Lumilux De Lux) or an equivalent from other manufacturers. These give more pleasant and natural color rendition than the type Cool White used since 1970. However, there seems little hope of a breakthrough in candling table design that would make nematodes or bones much easier to find.

Other methods for detection involve obtaining a picture of the fillet in some way or another. The use of film is hardly practical; a TV monitor is more appropriate if the picture is to be viewed by a human operator at all. A computer would simply observe the picture in its own memory. Either way, there is no reason to limit oneself to visible light. The list of possible detection methods now must include ultraviolet and infrared light, fluorescence, X-rays, and ultrasound. The task is then to select the most promising method and to determine the optimum wavelength and geometry.

On the other hand, irrespective of the method chosen, the picture may be sharpened by computer processing. The present chapter will only discuss detection by light, visible or infrared, and leave out picture processing.

Cost is always an important factor. We have, for example, found that worms can be detected by X-ray tomography, but the cost of such instruments is so high that they seem out of the question for the near future.

In order to clarify the aims of research into the problem at hand, it is better to have some idea of how a possible detection machine might be used as a part of a processing line in the factory. It is easy to suggest futuristic scenarios of robots picking worms and bones from fillets as they move past on high-speed conveyor belts. Such possibilities are too far in the future; a more realistic approach is to break the development into smaller steps, each with a reasonably close time goal.

A sensible first goal would be a machine for sorting fillets into defect-free, lightly contaminated, and heavily spoiled fillet lines. Only the lightly contaminated fillets would go on to ordinary candling.

To reach this goal, the only gap in our knowledge that needs to be filled by considerable new research is how to obtain good enough pictures, and how to process those pictures. Knowledge obtained as a result of such research will be useful in the later stages of development (even a robot must "see" a defect before removing it from a fillet). Furthermore, an understanding of the optical properties of fish flesh could result in new methods of quality control.

Preferably the same system should be able to detect several types of defects and act accordingly.

1.1 The Water Window

Probably the most difficult requirement is to be able to detect nematodes deeply embedded in a fillet that may be 5 cm thick. Other defects are mostly in the surface. A substantial fraction of the available light must survive travel through this thickness. About 80% of the flesh is water. It is reasonable to suppose that fish flesh absorbs a similar amount of light at each wavelength as water, if not more. Figure 1 shows the absorption vs. wavelength curve for fresh distilled water (4). The right vertical axis shows in cm the distance in which the light intensity falls to 10%. The lower limit of transmission that we can live with may be taken as 10%

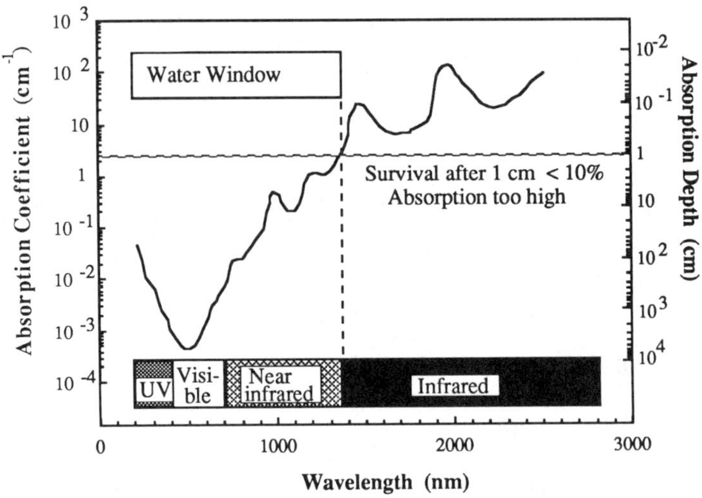

Figure 1 Absorption of light by pure water at different wavelengths. Water is opaque at wavelengths outside the "water window."

through 1 cm of fish (corresponding to absorption coefficient 2.3 cm^{-1}). This gives 1% light transmission through 2 cm and so on. Optical methods will be impractical if the transmission is less than this. The 10% line is indicated in Figure 1.

Wavelength of visible light is in the range 400–700 nm (nm = nanometer = 10^{-9} meter). Below 400 nm we have the ultraviolet (UV) range, and beyond 700 nm the near infrared. Far infrared or thermal radiation has even greater wavelength—3,000–30,000 nm (3–30 μm).

According to Figure 1, the 1300–2500 nm range is ruled out because the absorption is too high. The same applies to the range beyond 2500 nm (not shown) (5).

One conclusion of the very high absorption in the far infrared is that *the Hot Worm Method does not work.* This is the idea that nematodes could be selectively heated by, e.g., violet light and detected by their increased thermal radiation. It fails because the thermal (far infrared) radiation is very strongly absorbed by the fish flesh. This results in diffusive heat transport. Heat from a nematode spreads into a spherical volume of fish flesh, which is about 30 cc if the nematode is embedded 2 cm under the surface.

Figure 1 shows little absorption in the visible range (obviously, since water is reasonably transparent) and into the UV. It should be remembered that biological materials are sensitive to ultraviolet light.

Taking all this into account, it seems sensible to concentrate on the range 300–1300 nm. This window covers three times the visible range.

1.2 Previous Work

References to optical detection of nematodes in fish, published as well as unpublished, refer mostly to visible light. Some unpublished reports indicate especially high values of optical absorption in nematodes for violet light and into the UV. High expectations were based on this observation. But results have been disappointing despite considerable work. The reason was not clear. Power (1), on the other hand, concluded that a candling table illuminated with white or red light gives better results than blue light. Here we seemed to be faced with contradictions. Also at this time, the near infrared was practically unexplored.

There was clearly a need for a systematic investigation on the transport of light in fish flesh and in nematodes in the wavelength range 300–1300 nm.

2. CONCEPTS AND EXPERIMENTAL METHODS

A substantial proportion of visible light is transmitted through a fish fillet. Why then is the fillet not transparent? The reason is obviously *scattering* of the light by the fibers of the fish flesh. The direction and phase of a lightwave is changed in some random manner on its way through the fillet. Information coded into the light beam, e.g., an image of a nematode, will thus be smeared out. The general rule is for light scattering to be less at longer wavelengths.

In addition, some light is absorbed and converted into heat. The combined effect of scattering and absorption is *attenuation*. The fraction of the original light beam that is neither absorbed nor scattered on its way through the fillet is the *direct transmission*. The relation between these phenomena is frequently described by an extension of Beer's law:

$$T = C_1 \exp[-(K + S)D]$$

where T is the direct transmission calculated as the ratio of the undeviated beam to the incident beam. The parameter C_1 is that portion of the original beam that is not reflected or scattered back on the lamp side of the fillet, usually C_1 is in the range 0.5–1.0 for cod flesh. The thickness of the fillet or sample is D. The coefficients K and S are the absorption and scattering constants, respectively. In general both coefficients are wavelength dependent. On traversing a distance $1/(K + S)$, the direct beam is attenuated by 63%. The sum $(K + S)$ is the *attenuation coefficient.*

A different method is to measure *diffuse transmission,* i.e., to collect and measure all light that makes it through the sample. This corresponds to the equation:

$$T = C_2 \exp[-KD]$$

Measurements of T and D enable us to calculate the absorption K. Approximation of C_2 by unity corresponds to a neglect of reflected light. The resulting error is small and leads to slightly higher values of absorption.

The samples were pieces from cod fillets, suitable cut to fit into a cylindrical sample holder with transparent ends. Diameter of the incident beam was about 1/3 that of the sample holder. The holder fits into one port on an *integrating sphere* with 15 cm diameter and 98% diffusely reflecting inside surface. The sphere has three ports set at 0°, 90°, and 180°. The light detector fits into the 90° port. The transmission T is the ratio of signal from detector with sample in holder to signal without sample in holder. The geometry is shown in Figure 2. Transmission varies with wavelength.

In this manner, we have measured the absorption spectra (as opposed to attenuation spectra) of cod muscle and of nematodes (*Phocanema decipiens*).

For measurements of light scattering, the geometry shown in Figure 3 was used. A light detector with 5 mm aperture 92 mm from the sample surface was positioned in seven different posi-

$$\text{Diffuse Transmission} = \frac{\text{Light Signal with Sample}}{\text{Light Signal without Sample}}$$

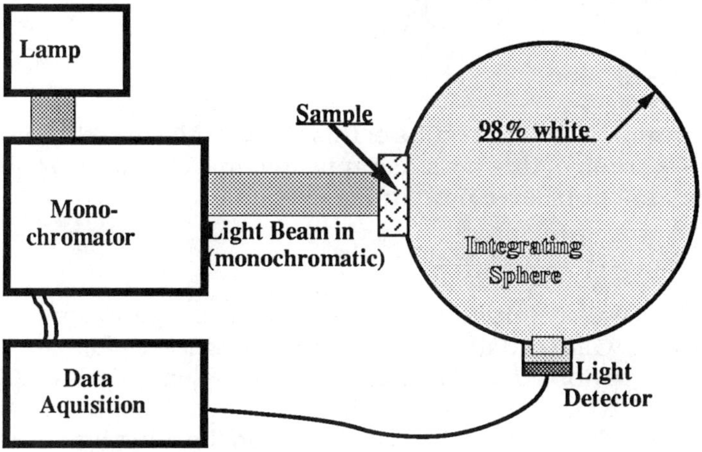

Figure 2 Setup for measurement of diffuse transmittance of biological tissue.

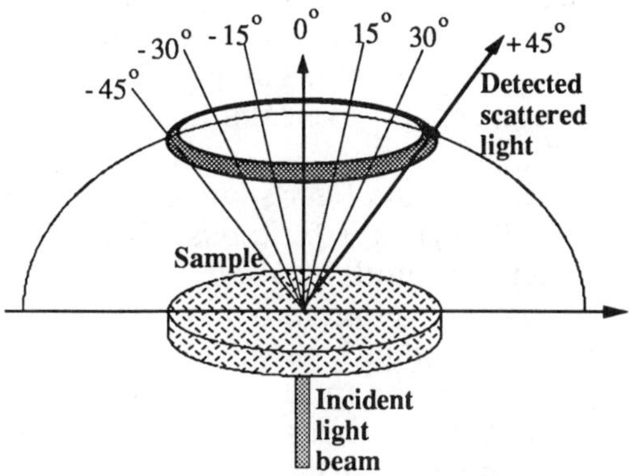

Figure 3 Geometry in measurements of scattered transmitted light as a function of angle and wavelength. A collimated beam from a monochromator enters the sample vertically from below. The transmitted light spreads out into the space above the sample. The detector can be moved in 15° steps along a circular arc centered at the point where the sample surface meets the incident beam axis. Distance of detector from sample center is about 20 times larger than the detector diameter.

tions in a sequence: +45°, +30°, +15°, 0°, −15°, −30°, and −45° from the incident beam axis.

3. RESULTS

We present below the results of this work for the first time in English. A portion of this work has appeared in Icelandic (6,7); the remainder is published now for the first time.

3.1 Optical Absorption

In Figure 4 transmission T is plotted in % as a function of wavelength for white cod muscle of several thicknesses (6).

It is obvious that very little light of wavelength below 400 nm or above 1400 nm is transmitted. Of course, T depends on sample thickness. The absorption coefficient K, on the other hand, is independent of thickness. It is a general parameter characterizing the material properties irrespective of geometry.

Figure 5 shows the absorption coefficient calculated from the transmission data in Figure 4. It was assumed that the constant C_2 = 1. Now the three curves coincide, indicating that assumptions in the equations hold. Other samples for which the results are not shown gave similar results. The absorption shows minima at 800–900 nm and at 1100 nm. The peaks just below 1000 nm and at 1200 nm correspond to peaks in absorption of water as shown in Figure 1.

The data in Figure 5 and Beer's law may be used to calculate the absorption in a fillet of given thickness. The effects of light scattering should be kept in mind, however.

Figure 6 shows absorption in nematodes, in brown fat from underneath cod skin, in slightly blood-stained cod muscle and in white muscle (same curve as in Fig. 5). Since this is absorption, the effects of different thicknesses have been eliminated. The results for nematodes confirm the high absorption in violet light previously observed. Note that at all wavelengths the absorption of the nematodes is at least twice that of cod.

The difference in absorption between white flesh, on one hand, and bloody flesh or fat, on the other hand, is mostly in the visible range. This is as expected from the brown or red appearance of these features. A blood clot would give a much more pro-

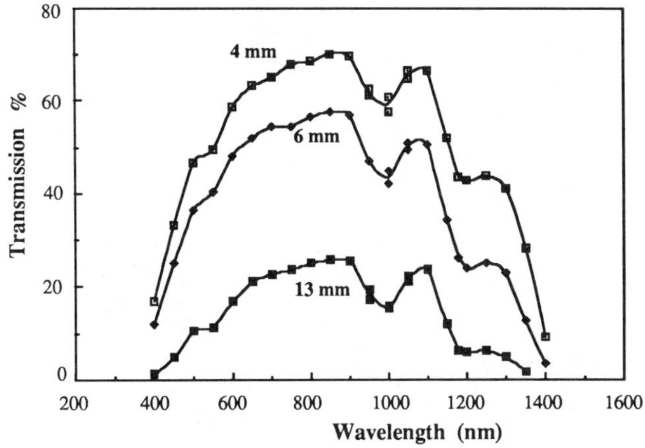

Figure 4 As measured optical transmission in cod flesh samples of three different thicknesses. Transmission depends both on sample thickness and wavelength.

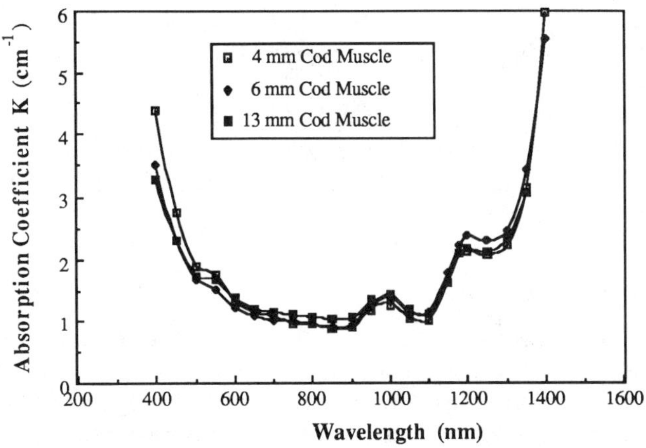

Figure 5 Absorption coefficient of cod flesh. Effects of sample thickness almost eliminated.

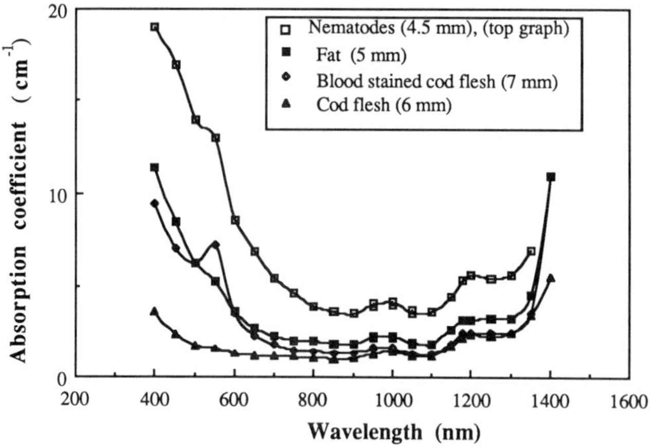

Figure 6 Optical absorption in cod worms, cod fat, and blood-stained cod muscle. Notice the high absorption at short wavelengths. Results from Figure 5 on pure cod are included for comparison.

nounced absorption peak than the slightly blood-stained muscle used here.

Looking at optical absorption on its own, it is clear that a substantial portion of 500–1200 nm light penetrates through a fillet. The difference in absorption between cod and worms is most marked in the 400–600 nm range. This, however, is in the visible range, where experience with candling tables indicates too low visibility of deeply located worms, even though those near the surface are easily detected.

The reason for this low visibility is scattering of light.

When choosing wavelength bands for an image system in fish processing, it is necessary to consider the combined effects of absorption and scattering.

3.2 Scattering

Observations through optical band filters of different wavelengths show that a worm on the surface stands out most clearly in violet to green light. This is in good agreement with Figure 6. However, similar observations, as well as photography, indicate something different if the worm is a few mm under the surface. In this case blue light shows a smeared out dark patch where yellow, red, and

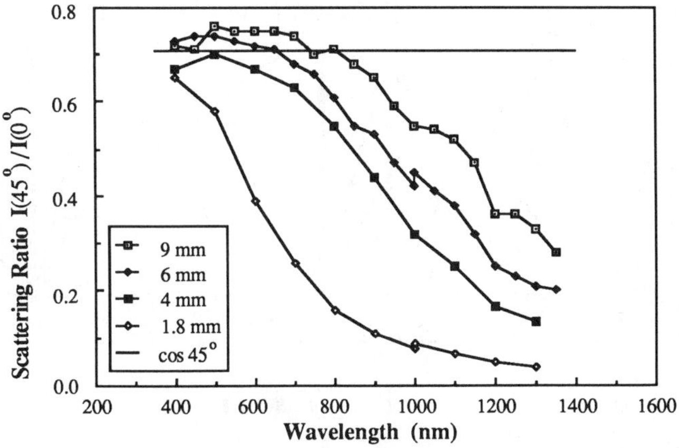

Figure 7 Wavelength dependence of scattering as measured in the setup from Figure 3. The proportion of light scattered into a small solid angle 45° from the incident beam direction is a measure of the amount of scattering. Totally scattering sample gives cosine distribution.

infrared give a faint but more recognizable coiled worm image. These observations support the results of H. E. Powers (1) that candling in red light gives better results than candling in blue light. *A logical conclusion is that blue light is scattered more in fish flesh than is red light.*

This is confirmed by direct measurements of scattering as a function of angle. Some results are indicated in Figures 7 and 8. Monochromatic vertical beam was incident on a horizontal sample from below. The rectangles represent the fish sample, and the length of the lines at ±15°, ±30°, and ±45° to the *vertical arrow* is proportional to the amount of light detected for these same angles. Each line has been normalized by the 0° measurement. The dotted circle indicates the limit for high scattering (all memory of original direction lost).

According to these results, blue-green light is almost completely scattered by a 4-mm-thick sample. Results on the same sample for 900–1300 nm show progressively less scattering as the wavelength increases. A sample of nematodes 2 mm thick, gives completely scattered beam at all wavelengths.

It is desirable to try to obtain a numerical measure of light scattering in cod flesh as a function of wavelength. This can be

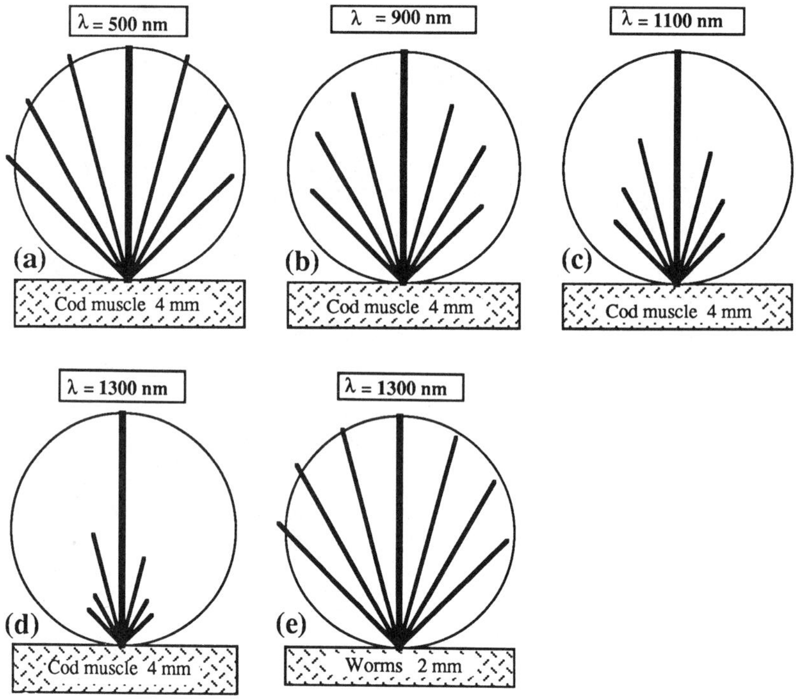

Figure 8 Angular distribution of scattered light at given wavelengths. The length of each radiation line is proportional to the scattering ratio in that direction. A beam of green light is totally scattered by a 4-mm-thick layer of cod flesh (a). Scattering in cod is shown to decrease with increasing wavelength (b)–(d). A 2-mm-thick sample of cod worms is totally scattered at all wavelengths (e).

done by comparison of results for samples of different thicknesses.

Figure 9 shows the ratio I(45°)/I(0°) as a function of sample thickness for nine different wavelengths. This data contains important information on visibility. The procedure used to extract this information is to go horizontally between neighboring curves for wavelengths λ_n and λ_{n+1} and find the sample thicknesses d_n and d_{n+1} that give the same scattering ratio. The ratio d_{n+1}/d_n then gives a measure of the *relative visibility* in the fish flesh at λ_{n+1} compared to the visibility at λ_n. Figure 10 plots the visibility at

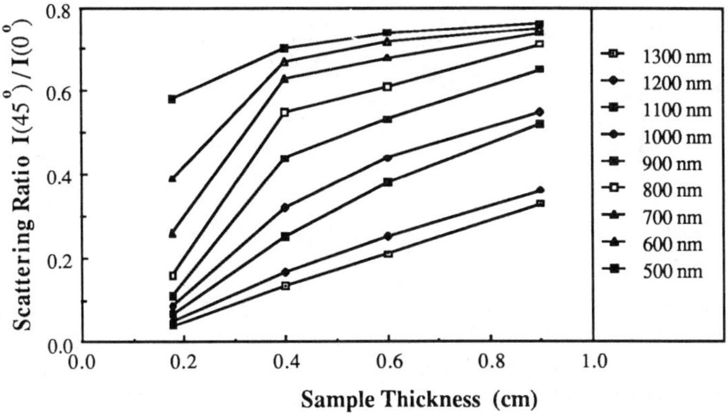

Figure 9 Scattering ratio—same data as in Figure 7 is now plotted vs. sample thickness.

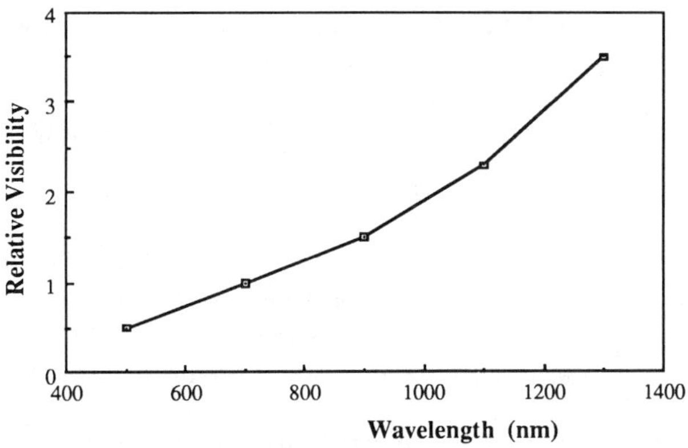

Figure 10 A rough measure of visibility through cod flesh is obtained from the graph in Figure 9 by working out ratios of thicknesses that give the same scattering ratio at different wavelengths.

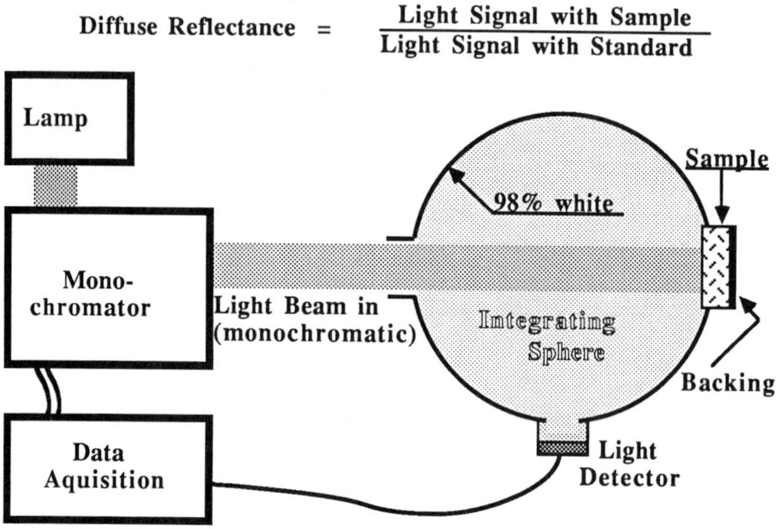

Figure 11 Setup for measurements of diffuse reflectance. The measured signal depends on absorption and scattering of the sample, and the absorption of the backing material. (See Appendix for details.)

eight wavelengths compared to the visibility at 700 nm, which has been arbitrarily set to unity. The ratios used are averages of results taken at three different values of I(45°)/I(0°). Clearly the visibility increases several times as we go from 500 to 1300 nm wavelength.

3.3 Diffuse Reflectance

This is another method of investigation that can give information on the K/S ratio (7). The results can be combined with data on K from diffuse transmission measurements to give the scattering coefficient S.

Alternatively, the diffuse reflectance for a thin sample (meaning SD $\leq$ 1) is measured with two different backing materials of known reflectances (see Fig. 11). Solving the so-called Kubelka-

Munk (8,9) equations for this situation, one obtains K and S separately (see Appendix).

Figure 12 gives results of the latter kind for a 5-mm-thick sample of cod flesh. The curves diverge above 500 nm. The backing has no effect for a 2-mm-thick sample of worms, however. Figure 13 shows the K/S ratios calculated from Figure 12. In Figure 14, K and S for cod have been separated by using the two curves in Figure 12. The values of K compare favorably with the results of Figure 5.

4. DISCUSSION

We are now in a position to quantify further the previous discussion on visibility. We adopt the following working definition of visibility:

$$S \times \text{Visibility} = 1$$

where S is the scattering coefficient as before. In Figure 15 we have plotted versus wavelength the product of S from Figure 14 and the relative visibility from Figure 10. This product is clearly independent of wavelength as expected. The relative visibility in Figure 11 happens to be only 26% higher than the absolute visibility 1/S.

In Table 1 we list the corrected absolute visibility at a few wavelengths.

Table 1 Absolute Visibility

Wavelength (nm)	Visibility (cm)
500	0.4
600	0.6
800	1.0
1000	1.5
1200	2.3
1300	2.8

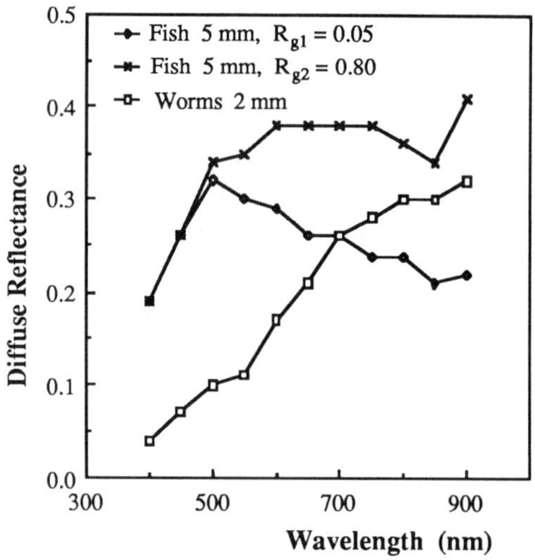

Figure 12 Diffuse reflectance of a sample of cod flesh with different backing materials. Notice that at short wavelengths the backing has no effect, no light survives two trips through the sample. Similar applies to a thinner sample of worms at all wavelengths.

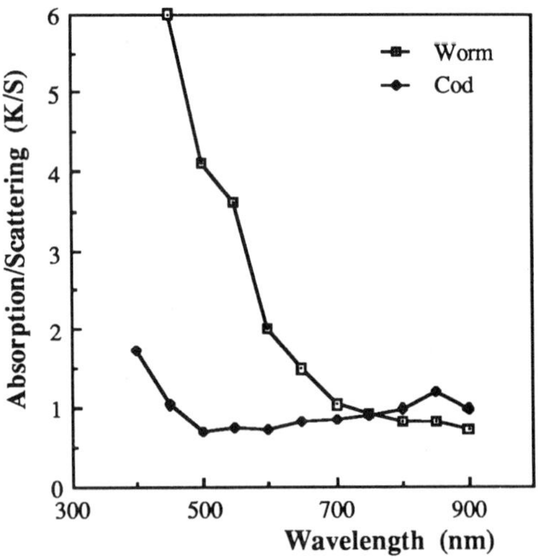

Figure 13 The ratio of absorption and scattering coefficients obtained from the data in Figure 12 by application of Kubelka-Munk theory.

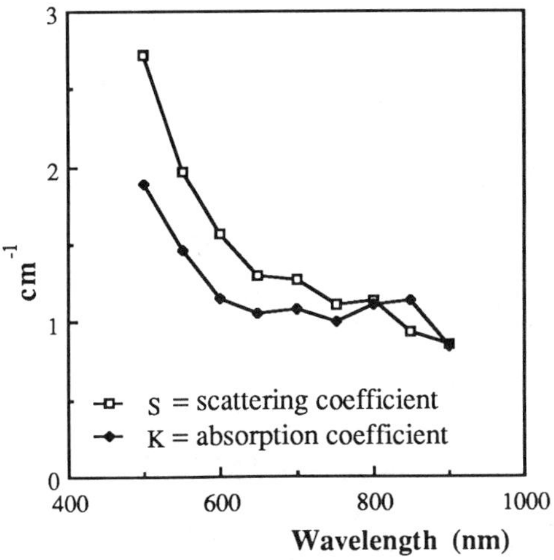

Figure 14 Scattering and absorption coefficients of cod can be obtained individually from Figure 12 at those wavelengths where reflectance depends on the backing. It is clear that cod flesh scatters light much more strongly at short wavelengths. Notice the good agreement in results for K found by this method and from transmission (Fig. 5).

It is quite clear that visibility grows to many times with increasing wavelength. Wavelengths longer than about 1300 nm are, however, ruled out by increasing absorption, since 80% of cod flesh is water.

We should also keep in mind the following points:

1. A camera can detect smaller differences in intensity than a human eye (contrast enhancement). This could double the visibility.
2. Image processing should increase visibility what ever the wavelength.

4.1 Fluorescence

It has been known for at least 40 years that fish bones and nematodes fluoresce under ultraviolet illumination (10). The maximum of the excitation spectrum for bones is near 340 nm and the emission peak at 390 nm extending into the visible range to about 440 nm (11). For nematodes the excitation spectrum peaks around

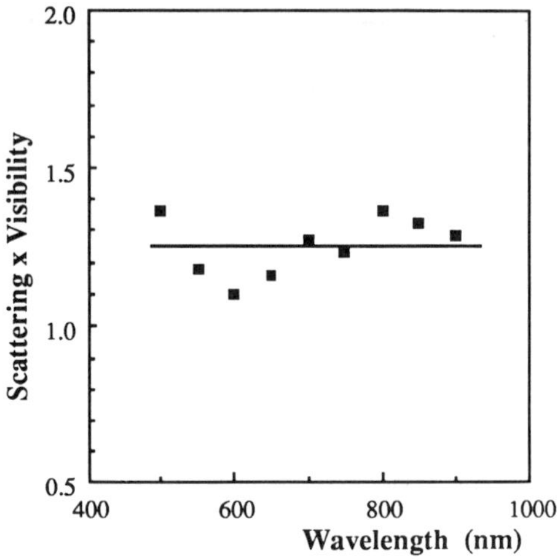

Figure 15 A plot showing the product of the scattering coefficient S in Figure 14 and relative visibility from Figure 10. This product is independent of wavelength within the experimental error. This is in agreement with the expectation that visibility ought to be inversely proportional to the scattering coefficient.

360 nm (12). The intensity and spectrum of the emission depends on previous treatment and perhaps even on the stage of development of individual larvae. Some nematodes may not fluoresce at all. Fish flesh also fluoresces in some cases (13), depending on previous treatment (freshness, freezing, etc.).

Light at the wavelengths involved in the fluorescence processes is, according to the results of the present work, strongly scattered. Absorption is also considerable. This leads to the conclusion that fluorescence methods are only useful for surface imaging. This is supported by the observation of Huss et al. (11) that the emission intensity decreases by a factor of 2 or more for each mm of flesh (corresponding to 0.1% surviving after 1 cm compared to 10%, used as a criteria in Figure 1).

4.2 Effects of Geometry

The above results clearly indicate the anisotropic nature of light scattering in fish flesh. It is a case of predominant forward scatter-

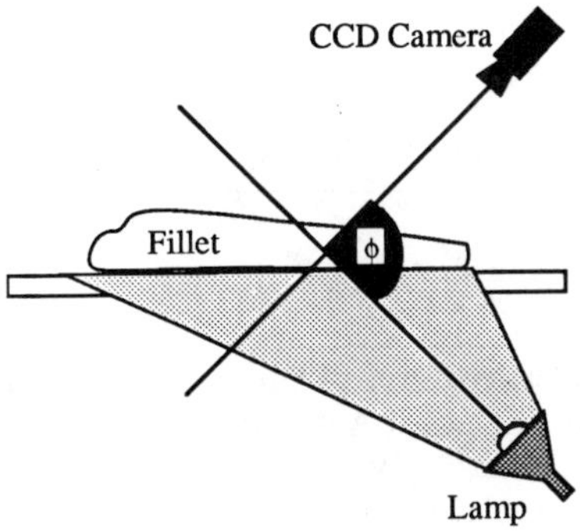

Figure 16 Lamp-fillet-camera geometry.

ing. The scattering from nematodes and other defects is much more isotropic.

An imaging system should be arranged so as to take advantage of this information if possible. Consider a lamp-fillet-camera system working in the 900–1300 nm range. The fillet rests on a transparent table or conveyor belt. The geometrical parameter of importance here is the angle ϕ between the direction of the light beam and the direction that the camera looks into (Fig. 16).

For $\phi = 180°$, when the camera looks directly into the beam, a defect should appear darker than the surrounding flesh. This is due to the combined effect of higher absorption of defect than of the flesh, and of more light scattering out of the direct beam on traversing the defect than flesh. The drawbacks of this geometry are that the image of the defect may be a small dip in a very bright background, which may saturate the camera. Also, if the defect is small and situated near the lampside of the fillet, some light will go around it precisely due to forward scattering in the fish flesh.

The $\phi = 90°$ geometry prevents the direct beam from entering the camera. The image should show light defects on a darker background. However, higher absorption of the defects has the opposite effect, i.e., tends to make the defect look darker. This absorption effect on the image will decrease with decreasing ϕ.

Backreflection from the fillet surface will in general make the $\phi = 0°$ geometry unattractive.

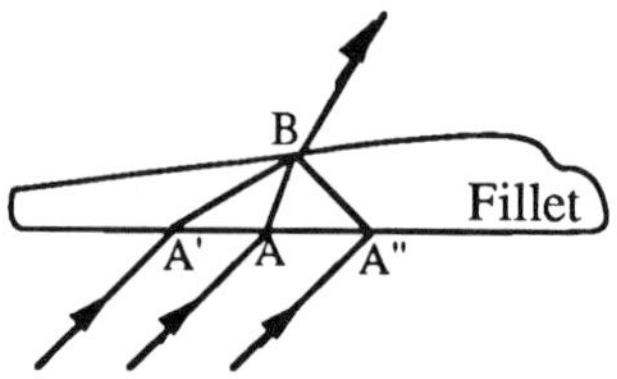

Figure 17 Potential advantage of laser scanning. A narrow laser beam is incident on a defect at point A in the surface of a fillet. The intensity coming from point B is strongly affected by optical properties of the defect. A wide beam is much less affected, on the other hand, because the measured light coming from point B may have entered at points such as A′ or A″. Notice also the refraction effect of the nonparallel surfaces.

4.3 Laser Scanning

This author is currently involved in experiments on laser scanning of cod fillets. This work is a collaboration with Þorsteinn J. Halldorsson at MBB in Munich who has published work on propagation of laser light in biological tissue (14,15) and Prof. Sigfûs Björnsson at the University of Iceland.

Discussion of results would be premature at this stage. However, the motivation for this work is closely connected to the present discussion.

What are the potential advantages of laser scanning for vision in the fish industry? This is best explained by a diagram (Fig. 17). Suppose we have a defect at point A near the lampside surface of a fillet. A camera or a detector will observe the attenuation due to this defect as a reduced intensity of the light coming from point B on the camera side of the fillet. However, some of the light coming from point B has been scattered around the defect at point A. Thus some of it may have entered the fillet at points A′ and A″. This light is completely unaffected by the defect at point A and therefore causes a reduced signal noise ratio. It should now be obvious that if no light enters at points other than A, the signal-noise ratio will be considerably higher. This may be accomplished with a narrow laser beam scanning the fillet point for point.

The disadvantages of laser scanning are complexity and high cost. Ideally one would like to scan synchronously with detector and laser beam. This is impractical, since, apart from other things, the surfaces of a fillet are not parallel to each other. This causes a transmitted laser beam to exit in a slightly different direction due to refraction at both surfaces. Perhaps a combination of a laser and CCD camera with a suitable signal processing can minimize the complexity and at the same time retain most of the advantages of synchronous scanning.

A recently demonstrated technology with great potential for imaging of biological materials some time in the future is optical ranging (16). This is in effect an optical radar that works by timing backscattered light from femtosecond laser pulses. Due to the enormous speed of light, submillimeter spatial resolution requires pulse length and time resolution better than 10^{-11} sec.

4.4 Optical Properties of Tissue from Other Fish Species

We have also applied the methods of diffuse transmission and diffuse reflectance measurements to other fish species. Haddock flesh gives similar results as cod flesh. Herring muscle shows considerably higher scattering than cod. This is presumed to be due to tiny fat globules dispersed in the flesh. Herring milt has extremely high scattering coefficient ≥ 50 cm^{-1} making it practically opaque. Herring roe scatters much less, particularly in the last weeks before spawning. An interesting effect is that a female herring that has been frozen and rethawed is considerably more transparent than the same herring when fresh. Greenland halibut muscle has less scattering effect than cod muscle. This is interesting, since, like herring, it contains a high proportion of fat. Apparently this is a more liquid or oily kind of fat.

APPENDIX

The Kubelka-Munk Scattering Model

Two channels: Up and down

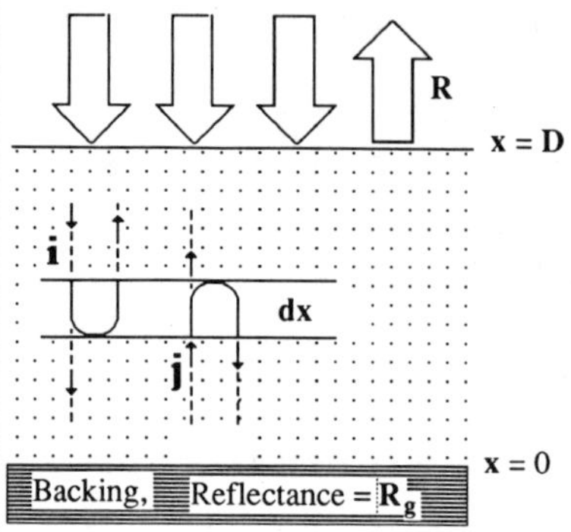

i = Intensity of downward light
j = Intensity of upward light

Down: $$\frac{di}{(-dx)} = -(K+S)\,i + S\,j$$

Up: $$\frac{dj}{dx} = -(K+S)\,j + S\,i$$

K is due to absorption,
S is due to scattering

$$j/i = E \;, \quad \frac{dE}{dx} = \frac{d(j/i)}{dx}$$

$$\Rightarrow \quad \frac{dE}{dx} = S - 2\,(K+S)\,E + S\,E^2$$

Integrate and solve for **E** = *Reflectance*

$$\int_0^D dx = \int_{E=R_g}^{E=R} \frac{dE}{[S - 2(K+S)E + SE^2]} \quad \Rightarrow$$

$$R = \frac{1 - R_g\,[A - B\coth(BSD)]}{A - R_g + B\coth(BSD)}$$

$$A = 1 + K/S, \qquad B = \sqrt{A^2 - 1}$$

This is the Kubelka-Munk equation

$$\lim_{BSD>>1} R = \frac{1 - R_g\,(A-B)}{A - R_g + B} = A - B = R_\infty$$

So that

$$R_\infty = 1 + \left(\frac{K}{S}\right) - \sqrt{\left[\left(\frac{K}{S}\right)^2 + 2\left(\frac{K}{S}\right)\right]}$$

gives the ratio **K/S** if the backing has no effect

$$\lim_{S\to 0} R = R_g\, e^{-2KD}$$

i.e., case of no scattering

To obtain **K** and **S** separately, one must either find **K** from measurements of the *diffuse transmission*, or use a *thin sample* and measure **R** for two different values of **Rg** , i.e., for two different backing materials.

$$E_1 = \frac{1 - R_{g1}\,(A - C)}{A - R_{g1} + C}$$

$$E_2 = \frac{1 - R_{g2}\,(A - C)}{A - R_{g2} + C}$$

Here **C** = **B** coth(**BSD**), and the only unknowns are **A** and **C**, which depend on **K** and **S**.

Therefore, inversion of these equations finally gives **K** and **S**.

REFERENCES

1. Power, H. E. 1958. *J. Fisheries Board Canada* 15(4):537–542.
2. Àgùstsson, D. 1970. Unpublished Report to the Icelandic Fisheries Laboratories.
3. Pétursson, J. 1987. Selection of Fluorescent Lights for Candling Tables in Freezing Plants, *Report to The Icelandic Freezing Plants.*
4. Neumann, G., and W. J. Pierson Jr. 1966. *Principles of Physical Oceanography*, Prentice-Hall, Englewood Cliffs, NJ.
5. U.S. Office of Naval Research, *The Infrared Handbook*, pp. 3–107.
6. Pétursson, J. 1984. Fish Optics and Nematode Detection, *Fiskvinnslan*, Professional Journal of the Fish Industry No. 1, pp. 11–15.
7. Pétursson, J. 1984. *Physics Research in Iceland*, Proc. of the 2nd Conf. of the Icelandic Physical Society, Munaðarnes 29–30 Sept., pp. 145–152.
8. Kubelka, P., and F. Munk. 1931. *Z. Tech. Phys.* 12:593.
9. Allen, E. 1980. In *Optical Radiation Measurements*, Vo. 2, F. Grum and C. J. Bartelson, (eds.). Academic Press, NY.
10. Sinnhuber, R. O., and D. K. Law. 1949. Oregon Fish. Commission Research Brief 2(1):7–9.
11. Huss, H. H., P. Sigsgaard, and S. A. Jensen. 1985. *J. Food Protection* 48(5):393.
12. Pippy, J. H. C. 1970. *J. Fisheries Res. Board Canada* 27(5):963.
13. Davis, H. K. 1982. *J. Sci. Food Agric.* 33:1135.
14. Halldorsson, T., and J. Langerholc. 1978. *Appl. Opt.* 17:3948.
15. Langerholc, J. 1982. *Appl. Opt.* 21:1593.
16. Fujimoto, J. G., et al. 1986. *Opt. Lett.* 11(3):150.

4

Optical Properties of Fish Classes

KEPA BENGOETXEA *Dow Chemical Company, Tarragona, Spain*

1. Introduction
2. Skin
3. Flesh
4. Survey
Reference

1. INTRODUCTION

There are about 20,000 different living species of fish in the world, but we will consider in this chapter only those of economic importance. They are found in the group of bony fishes (this group includes by far the majority of living fishes) such as the herring-like fish, cod group fish, mackerel family fish, and flat fish.

Our purpose is to review the optical properties of skin and flesh of different species in order to build the illumination part of a vision system.

2. SKIN

The outer skin or epidermis of a fish secretes a large amount of mucus which serves partly to reduce the frictional resistance of the water and partly as a barrier to parasites (1).

Furthermore, the body of a bony fish is covered more or less completely by scales (1). The distribution and possible existence of the scales should be considered because the scales can give the same effect as a mirror. In several species of the tuna group, only the breast region has an armor of large scales, while the rest of the body and the tail are naked. In other species, the skin is furnished with large bony thorns or is armored with bony shields. Scales are lacking on very young fish, and are only developed when the fish has reached a certain length. After that the scales grow with the fish and they form rings.

Because of these two characteristics—mucus and scales—fish should be considered as a shiny object for vision applications.

It is interesting to consider the difference between the fresh and frozen fish. Due to the cold, the content of water in the skin decreases, and the fish becomes drier and more matter.

3. FLESH

Fish flesh has a relatively poor supply of blood and is usually white. However, fast-swimming fishes have a more rapid metabolism and a better blood supply to their muscles (1). Their flesh therefore has dark areas or is reddish (e.g., in the tuna).

The color of the flesh will change due to the cold too, especially in the cod and flat fish group. The flesh becomes yellowish, so discrimination of the blood spots would become more difficult. It would be the same for bones and parasites using optical light.

4. SURVEY

Table 1 shows for each fish specimen optical characteristics, such as skin color, existence of scales, and meat color.

According to the table, special attention should be paid to:

- The discrimination of blood spots, bone, and parasites from the flesh in those species with white flesh under direct illumination if they have been frozen.
- The glinting for fresh fish because of the reflection caused by the mucus.
- The glinting in those species which have special remarks in the scales column because of the reflection of the scales.
- The discrimination of blood spots and black skin from the flesh in those species with reddish flesh.
- The measures to achieve discrimination between the background and dark top skin, as well as the white bottom skin in flat fish species under direct illumination, for length or shape estimation.
- The polarization of the light in order to discriminate and segment the fish from the background, especially in the belly part (silvery color) under direct illumination.
- The color discrimination of the flesh from the background in redfish due its pale color under direct illumination.
- The discrimination of the flesh from the background in megrim under backlighting because of its transparency.

Table 1 Optical Properties of Fish Species

Fish species	Color	Scales	Flesh color
Herring (*Culpea harengus*)	dark green on the back, silvery belly	covered along the midline on each side	reddish
Pilchard (*Sardinia pilchardus*)	dark blue on the back, silvery belly	30 scales along the midline on each side	reddish
Salmon (*Salmo solar*)	dark blue on the back, silvery belly	11–15 scales between the adipose fin and the lateral line	white
Trout (*Salmo trutta*)	gray with black spots, silvery belly	14–19 scales between the adipose fin and the lateral line	white
Saithe (*Pollashius villosus*)	black to gray in the back, silvery belly	no remarks	white
Haddock (*Melanogrammus aeglefinus*)	grayish on the back silvery belly	no remarks	white
Cod (*Gadus morhus*)	reddish on algal zone, greenish among eel–grass, pale gray on sandy bottom, silvery belly	no remarks	white
Mackerel (*Scomber scombrus*)	dark green/blue, silvery belly	armor of scales in breast region	reddish
Tuna (*Thunnus thynnus*)	dark blue on the back, silvery belly	armor of scales in breast region	reddish
Catfish (*Anarhichas minor*)	brownish	no remarks	white

Table 1 (Continued)

Fish species	Color	Scales	Flesh color
Red fish (*Sebastes marinus*)	pale reddish	16–21 scales between the rear end of the anal fin and the lateral line	white
Turbot (*Scophthalmus maximus*)	brownish back, white	no remarks	white
Flounder (*Platichythys flesus*)	grayish/brownish back, white bottom	no remarks	white
Megrim (*Lepidorhombus whiffiagonis*)	clear brownish back, white bottom (trans–lucent)	no remarks	white

REFERENCE

1. Muus, B. J., and P. Dahlström. 1974. *Collins Guide to the Sea Fishes of Britain and North-Western Europe.* Collins, London.

5

Parasitic Nematodes in Commercially Important Fish

ERLINGUR HAUKSSON *Icelandic Fisheries Laboratories, Reykjavik, Iceland*

♦♦♦

1. Introduction
2. Parasitic Nematodes in Marine Fishes
    2.1 Taxonomy and Life Cycles
    2.2 Anatomy
    2.3 Physiology
    2.4 Chemical Information
    2.5 Effects on Hosts
3. The "Nematode Problem" in the North Atlantic
    3.1 Distribution and Magnitude
    3.2 Biology of the Most Important Species
    3.3 Solutions to the Nematode Problem
4. Nematodes and Fish Processing
    4.1 Migration of Nematodes in Dead Fish
    4.2 Efficiency of Candling in Detecting Nematodes
    4.3 Distribution of Nematodes in Fish Fillets

5. Public Health Aspects
  5.1 Anisakiasis
  5.2 Laws and Regulations Concerning Anisakiasis
  5.3 Reactions of the Public to Nematodes in Fish Products
6. Concluding Remarks
  References

♦♦

1. INTRODUCTION

This chapter presents the most important biological information on nematodes in commercially important fish species and the problems they cause for the fishing industry

In Iceland and other fishing countries of the North Atlantic, nematodes [codworms; sealworms (*Pseudoterranova* (= *Phocanema*) *decipiens*) and herringworms; whaleworms (*Anisakis simplex*)] in commercial fish have been a problem for several decades.

A solution to the nematode problem, which would be the total cleanness of fish products in regard to nematodes, could theoretically be obtained by physical or biological means, or both hand in hand.

2. PARASITIC NEMATODES IN MARINE FISHES

2.1 Taxonomy and Life Cycles

Anisakids are among the most common nematodes of marine fishes. They affect fish by causing pathological symptoms and possibly mortalities, they reduce the commercial value of fish, and they may infect man, producing harmful effects. Larval anisakids in fish belong mainly to the genera *Hysterothylacium*, *Contracaecum*, *Anisakis*, *Porrocaecum*, *Phocanema*, *Raphidascaris*, and *Terranova* (1). There may be several hundred larvae in one fish, and many fish species act as hosts.

The development of parasitic nematodes includes four larval stages and one adult stage. In many species, the stage II larva is found in the sheath of the stage I larva within the eggshell. In the intermediate hosts, this stage hatches and molts into the stage III larva. It is either free-living or encapsulated in the stomach wall, on the surface of, in, or between organs of the body cavity, or in the skeletal muscle of the fish. The stage III larva may repeatedly be taken up by transport hosts that eat the previous host. When entering the stomach of a suitable final host, this stage molts into the stage IV larva and soon after into the female or male adult nematode.

2.2 Anatomy

Nematodes (= roundworms) are thin, elongated, round worms without segmentation. Because of their uniform shape, they can easily be distinguished from other worm groups, but species determination is complicated. Nematodes have a complete digestive tract with mouth, esophagus, intestine, and anus. Main criteria for species determination are size, fine structure of head and tail, position of excretory pore, and structure of the transitional area between esophagus and intestines. Adult males carry extrudable spicula near the tail end and, in some species, a characteristic sucker (2).

The nematode cuticle is considerably more complex than that of other related phyla of worms. It consists of secreted collagen, which appears to be unique to nematodes, and it is organized within three main layers, a cuticle, an epidermis, and a muscle layer (3).

2.3 Physiology

Nematodes respond to chemical, mechanical, photic, and thermal stimulation. There are less clear reports of orientation to gravity and electrical potentials.

In connection with each change between larval stages, there is molting. The old cuticle is shed and a new one created. It is not certain what factors govern the initiation of molting, but neurosecretory granules have been shown to be associated with secretory cells in the sealworm at the time of molting, and it has been suggested that molting in nematodes may be controlled by a similar mechanism to ecdysis in insects. Insect juvenile hormone and its mimic, farnesyl methyl ether (FME), have been used to initiate normal molting experimentally, but whether these hormones are actually involved within nematodes naturally has still to be resolved (4).

2.4 Chemical Information

Nematodes have a very similar chemical content as flesh of cod, but are a little richer in carbon and copper. They are even more nutritive than fish. Sealworm larvae 43.5 mm in length weigh about 30 mg (5). They smell a little, which can be attributed to the constituents pentanol-3 and pentanone-3 (6).

2.5 Effects on Hosts

Herringworm (Anisakis simplex) Larvae

Infection of fish with larval *Anisakis* leads to marked tissue reactions. Larvae which have migrated from the stomach into the body cavity of fish gradually coil up and the fish host forms a connective tissue capsule around them. There is little conclusive evidence that herringworm larvae are serious pathogens of fish, although they have been observed to damage the liver of fish when infections are heavy. The fish muscles are apparently least effected (1).

Sealworm Larvae

Some histopathological findings on the infection with sealworms have been reported. Larval sealworms are, however, most frequent in the musculature, and do not infect organs of the body cavity to any degree. Most larvae are found in capsules of connective tissue of host origin. The connective tissue close to the nematode is often denser, forming a sheathlike structure. Sheaths may show degenerative process, as may the nematodes, leading to a shrinking of the capsules and nematodes which appear brown in color. In sea scorpion (*Myoxocephalus quadricornis*), sealworms have been observed to damage the liver (1).

Some evidence shows that cods with many nematodes (mostly sealworms) in the musculature are thinner than those cods which have few or none. This results in less flesh on the bones and a decrease in filleting efficiency at the fish processing plants, at an average of about 0.5% for each nematode (7).

3. THE "NEMATODE PROBLEM" IN THE NORTH ATLANTIC

3.1 Distribution and Magnitude

The infection level of sealworm in cod seems to be rather similar in Canada and Iceland, but is on the average much lower in British and Norwegian waters. Infestation levels of fish with herringworm are, on the other hand, much higher in the North Sea and off the coast of northern Norway than in the sea around Iceland and Greenland (8).

There has been a dramatic increase in sealworm infection levels in ground-fish from the Scotian shelf (eastern Canada) in recent years (9, 10). Such an increase, both in prevalence and intensity, has also been reported for cod in Iceland and in local coastal cod stocks in Norway (11). In Scottish waters (England), infection levels had increased noticeably after the turn of the century (12,13), although in some cases the increase is probably mainly due to herringworm. In the 1970s, sealworm infection levels in cod stabilized or even decreased (14).

Such reduction or stabilizing of infection seems also to have happened, at lease, in some coastal areas of Iceland the last years, where young cod is concerned. The herringworm infection levels have, on the other hand, increased in cod (15).

Cod is hardly mentioned as a source of human parasites, as is herring, and the larvae are removed from fish fillets if only for cosmetic reasons. The expenditures involved are great. They comprise the costs for additional labor and training of workers involved in candling cod as well as other infected ground-fish and flatfish species. Extra cost is also caused by a reduction in quality of the fillets if they have to be sliced or turned into fish flour as occurs in the case of heavy infestation.

In North Sea waters, herringworm infection in pelagic fish species, notably herring, is causing great concern at the moment. Although estimates of the commercial impact of the public response to a German television program, which showed live nematodes in fish, are not available, they are likely to have been very high. The costs include additional processing and job losses in Germany and other EEC countries, as the customer demand for herring and other fish dropped dramatically. The market has more or less stabilized again, but such lapses can be expected to recur as the public becomes more and more concerned with nematodes in fish.

3.2 Biology of the Most Important Species

Sealworm (Cod-worm)

Eggs of sealworms sink, and emerging ensheathed larvae attach themselves by their caudal extremities to the substrata. Survival of ensheathed larvae in seawater is temperature dependent, ranging from 6 days at 17°C to 140 days at 5°C. At 20°C, larvae die within 24 to 48 h; at 0°C they survived for 90–120 days. Larvae were experimentally successfully fed to benthic copepods. Larvae have also been found to naturally infect marine isopods, mysids, and decapods. Fish could not be directly infected with larvae from eggs, and the smallest nematodes in naturally infected fish are 4–5 mm long, which is further evidence for that invertebrates act as obligatory first intermediate hosts. Final hosts are seals. In the Atlantic, grey seals and common seals are significantly parasitized by *P. decipiens*, and pinnipeds have been shown to be the main hosts in the northern Pacific (1).

Herringworm (Whaleworm)

The final hosts are cetaceans, mostly toothed whales. The eggs leave the host in feces and develop at temperatures ranging from 2 to 27°C. At 2°C they hatch within 40 days. Larvae are eaten by various krill species (*Euphausiacea*), in which exsheathment occurs within 8 days. Molting to the third larval stage occurs before they reach 6 mm in length. Larvae longer than 19 mm can survive in fish and squid when ingested and penetrate through the wall of the digestive tract. Larvae can be repeatedly passed on to the other squid or fish without further molts. Gradually, larvae are concentrated in several large predatory fish species (1). The herringworm larvae get to the final host when the infected fish is eaten by whales.

3.3 Solutions to the Nematode Problem

Physical Solutions

In Iceland, it is a long time since scientists as well as laymen started trying to solve the nematode problem. The first experiments aimed at getting the nematode to move out of the fish fillets on its own. Several means were tried, such as smoking, blowing hot air over the fillets, cooling, direct electrical current, and keeping fish with nematodes in a CO_2, nitrogen, and oxygen mixture. Only smoking and blowing of hot air over the fillets had any effects, but not anything near what was hoped for. Only a small portion of nematodes exited, and this happened very slowly (16).

The only reasonable solution to the problem so far is inspecting the fish fillets on a candling table and picking the nematodes out, or cutting away heavily infected areas of the fish, such as the flaps and napes. In Iceland, fish-processing plants started in general using candling in 1960–65, and have used it since. This method involves much labor and the cutting to get the nematodes out damages the fish, which reduces its value on the market. Also it is far from being 100% effective, in regard to finding all nematodes. Much work is now being carried out to improve the efficiency of candling, and find better methods for detecting nematodes in the fillets.

Chemical Solutions

Solving the nematode problem by chemical means has been thought of, but not tried, as far as is known. Stimulating hormones look promising for getting the nematodes to move. Also, detecting the nematodes by the odor they give off may be possible. It is, however, of common knowledge that using drugs, hormones, and other chemicals in food production is a hazardous business, in light of negative reactions of consumers.

Biological Solutions

As described earlier, the final hosts of sealworm are seals, and the final hosts of herringworm are whales. At least where the sealworm is concerned, its geographical distribution and abundance in fish is highly correlated to the distribution and number of seals (9). By reducing the number of final hosts (seals and whales), it should be a possibility to stabilize the abundance of nematodes in commercial fish at lower levels than at present is the case. How great a reduction of the number of final hosts is necessary to decrease the nematode burden in fish is not exactly known.

4. NEMATODES AND FISH PROCESSING

4.1 Migration of Nematodes in Dead Fish

Experience and experiments indicate that if a whole fish is left lying for a considerable time, some of the nematodes in it start creeping around. When the intestines and organs in the belly spoil, nematodes start to come out of the fish through the gill openings and anus. Some of the nematodes may even creep into the flaps and increase the nematode burden in the fish by doing so. This is especially significant for herringworm and herring (17).

The optimum temperature for sealworms in fish is in the range of −2.5 to 7.5°C (18). Therefore, in properly kept fresh fish, the majority of nematodes are alive. They do not, however, move about much, but are in a sense prisoners in the capsules of the connective tissue. Only after the fish starts to spoil do some of the nematodes move. Freezing and salting the fish, therefore, prevents movement of the nematodes. The author has, for example, shown experimentally that during salting, from the beginning to the end of the salting process, the distribution of the nematodes in the fish does not change.

Recent experiments on cod in cooled seawater and freshwater tanks, carried out by the author, indicate that the nematode burden of the fillets did not change after the fish had been kept there for a week. And no change of nematodes in fillets was noticed in cods kept in fresh water at 15°C. It also did not make any difference whether the fish had been gutted or not. Quite a number of nematodes did creep out of the fish, especially in the whole cods, and were collected in plastic bags which surrounded each fish. But those nematodes were of species which are most common in the visceral organs of cod like *Contracaecum*, *Hysterothylacium*, and *Anisakis*, and not sealworms. Therefore, the opinion of many fish processors in Iceland,—that if cod is kept in cold seawater tanks for awhile, the nematodes will creep out of the fillets and out of the fish or into the liver—does not hold. And the nematode problem cannot be solved in that way (19).

Table 1 Number of Nematodes Hidden in the Flesh of Fully Processed, Salted Cod, when No Nematodes are Detected by Candling

Number of nematodes	Number of fishes	Frequency of fishes	Cumulative frequency of fishes
0	27	28.1	28.1
1	11	11.5	39.6
2	17	17.7	57.3
3	12	12.5	69.8
4	10	10.4	80.2
5	5	5.2	85.4
6	2	2.1	87.5
7	6	6.3	93.8
8	1	1.0	94.8
9	2	2.1	96.9
>10	3	3.1	100.0
Total	96	100.0	

4.2 Efficiency of Candling in Detecting Nematodes

Detecting nematodes in skinless cod fillets is far from being 100% effective, and even with the best available light conditions, efficiency higher than 75% can hardly be expected (20).

Detecting nematodes in unskinned cod fillets, as has to be done now in the production of salted fish, is of course even less effective, and studies carried out by the author show this to be only about 25% efficient (Table 1). This may be a conservative figure because fully processed salted fishes were used in the study, and in the plants nematodes are taken out sooner in the salting process. However, it is very unlikely that the figure of efficiency reaches 50%.

The efficiency of candling can also depend on fish species, nematode species, fish conditions, and other factors. Fishing grounds can also play a part in efficiency of candling, probably through difference in fish texture of the fish from different grounds. Proper bleeding of fishes is also very important. In cods made to bleed alive, nematodes are more easily detected by can-

Table 2 Percentage Distribution of Nematodes in Flesh of Cod

A. Sealworm (number of nematodes = 1074)

Parts	Flaps	Trunk	Ventral tail	Dorsal tail	Total
Right	15.4	14.0	6.6	4.6	40.6
Left	23.9	15.2	12.7	7.6	59.4
Total	39.3	29.2	19.3	12.2	100.0

B. Herringworm (number of nematodes = 974)

Parts	Flaps	Trunk	Ventral tail	Dorsal tail	Total
Right	43.0	2.3	1.5	0.0	46.8
Left	50.5	1.1	1.3	0.3	53.2
Total	93.5	3.4	2.8	0.3	100.0

dling than in cod which was already dead when gutted and made to bleed (21). This difference in efficiency can amount to 15%. There are also human factors involved. Proper working conditions, without noise and with periodic breaks in working at a candling table, proper training and good eyesight all increase the candling efficiency.

4.3 Distribution of Nematodes in Fish Fillets

Investigations on distribution of nematodes in cod indicate that nematodes are more abundant in the left than in the right fillet. Distribution of sealworms is very uneven between parts of the fillets, the left flap holding 24% of all sealworms, but dorsal muscles in the right side of the tail only 5% (Table 2).

Distribution of the herringworm in cod is even more skewed, in relation to parts of the fillets. About 51% are found in the left flap, in the right flap about 43%, so about 94% of all the herringworms are located in the flaps (Table 2). This is a much higher percentage than in the case of the sealworm (22). This results in a very clumped distribution of the nematodes in the lower ventral side of the flaps, as indicated in Figure 1.

```
0 0 0 1 0 0 1 1 0 0 0 0
0 0 0 0 1 1 0 1 1 0 0 0 0 0 0 0 0
1 0 0 0 0 2 3 0 1 1 0 1 0 2 0 0 0 0 0 1 0
0 1 1 2 0 0 0 3 0 2 0 0 1 1 1 1 2 0 2 0 0 1 0 1 0 0 0
0 0 0 0 3 0 0 0 2 3 5 3 3 2 2 0 1 2 0 0 1 0 0 0 1 0 0 0 0 0 0 0 0
0 1 1 2 1 1 1 2 2 2 1 0 2 0 0 4 1 0 1 0 0 0 1 0 0 0 0 0 0 0 0 0
1 0 8 8 8 9 4 2 2 1 0 1 0 1 2 0 0 3 0 1 0 0 0 0 0 1 0 0 0 0 0 0
0 0 0 6 10 12 3 7 4 4 3 2 4 1 3 1 3 0 1 0 1 1 1 0 0 0 0 0 0
0 0 2 1 17 15 9 12 12 6 6 5 2 7 1 5 2 2 1 1 1 0 0 0 0 0
0 0 0 2 9 12 11 17 11 11 9 11 11 13 2 4 2 0 0 0 0 0 0
0 2 0 0 5 3 10 14 15 13 8 11 8 1 0 1 0 0 0
0 0 0 0 3 5 7 9 10 16 6 7 2 0 0
0 0 0 1 1 2 2 0 3 2 5
0 0 1 0 0 2 1 1 1 1
```

```
1 0 0 2 2 2 1 0 2 0
1 0 1 0 3 1 2 5 5 2 5
3 1 2 2 5 6 13 13 19 18 27 33 11 3 0
1 0 2 11 11 14 23 28 31 27 33 57 43 31 21 5 1 0 0
1 1 6 9 9 16 17 24 13 23 34 29 29 26 18 2 4 2 0 1 0 0 0
0 0 13 14 11 7 8 17 12 18 17 16 20 21 7 4 5 2 2 1 1 0 1 0 0 0
0 1 3 7 6 2 4 3 9 7 6 16 6 5 2 5 4 0 1 1 1 1 0 0 0 0 0 0 0
1 1 0 2 2 5 2 2 4 1 2 3 2 2 2 2 1 0 0 1 2 0 0 1 0 0 0 0 0 0 0
0 0 1 0 1 2 1 2 2 1 2 2 1 1 1 0 1 1 0 1 1 0 2 1 1 0 0 0 0 0 0 0
0 0 1 0 0 1 2 0 2 1 2 1 4 2 0 2 0 1 0 0 0 0 0 1 1 0 1 0 0 0 0 0 0
0 0 0 2 1 1 1 2 2 2 3 0 3 2 0 0 0 1 1 0 0 1 0 1 1 0 1
0 0 0 1 2 0 0 1 3 2 0 2 1 1 0 0 0 0 0 0 0
0 0 0 0 0 1 1 0 0 1 0 0 1 0 0 0 0
0 0 0 0 0 1 0 0 0 0 0 0
```

Figure 1 Mean distribution of nematodes in the right (top) and left (bottom) fillets of cod. All nematodes taken together from 15 right and 15 left fillets, which were especially chosen because of heavy nematode burden. Each integer represents the total number of nematodes found in a core of the fillets of an area of 1 cm^2.

Table 3 Depth Distribution of Nematodes in Cod Fillets[a]

Depth	0–1 cm	1–2 cm	2–3 cm	Total
Right	594	23	4	621
%	95.7	3.7	0.6	
Left	1145	55	9	1209
%	94.7	4.6	0.7	

[a]Measured from the inner surface of the fillets, down into the fish flesh, with the fillets lying on a candling table skinside down.

The depth distribution of the nematodes is given in Table 3. About 95% of them are within the first 1-cm-thick layer of the inner side of the fillet, and should therefore be detectable by candling, because, according to investigations, those nematodes are detected which lie closer than 6–8 mm to the surface of the fillets (23).

5. PUBLIC HEALTH ASPECTS

5.1 Anisakiasis

Before 1955, it was commonly believed that the herringworm was totally harmless. But that year for the first time it was medically tied to a not so uncommon illness in the Netherlands. All patients had consumed a popular gourmet dish called *groene haaring* in the Netherlands. This dish is made of very lightly salted herring, which is consumed right after salting. In effect the fish is raw and fresh, and any nematode larvae hidden in it get into the stomach of the consumer, with anisakiasis (a name given to this disease after its cause became known) as a result. After anisakiasis had been described in the Netherlands, the Japanese authorities awakened. There this illness had been quite common for a long time without people knowing what caused it.

In the light of the fact that anisakiasis only occurs when people consume raw fish with herringworm larvae, it depends totally on diet and cooking procedures whether people get the disease or not. In countries like Iceland, where it is customary to boil, fry, or heavily salt fish before consumption, nematodes in fish are not a great public health problem.

5.2 Laws and Regulations Concerning Anisakiasis

Authorities in the Netherlands were the first to tackle the problem of anisakiasis, and in accordance with results of research on nematodes in fish made laws and regulations for the fish industry to follow so that infection of consumers with nematode larva should be impossible. According to these regulations it is forbidden by law in the Netherlands to sell improperly processed herring and mackerel, and products from herring and mackerel have to be frozen, heavily salted, temperature treated, or so strongly marinated that is certain that all nematodes have been killed.

In other European countries, a growing concern about nematodes in fish has occurred in recent years. This has manifested itself in inter-European cooperation in making rules regarding fishing and fish processing. The Commission of the European Communities has laid down drafts for health conditions concerning nematodes for the marketing of fish and fish products. In the draft it is strongly declared that fish may not be visibly infested with parasites, and that fish products ready for consumption may not contain live parasites.

The number one rule to follow for anyone to avoid getting anisakiasis is to not consume raw fish or raw fish products.

5.3 Reactions of the Public to Nematodes in Fish Products

According to the German media, nematodes in Danish herring triggered the strong reaction of the German people to nematodes in 1987. But a subjective television program surely did help. The German market collapsed to one third of the usual sales volume. The public showed an enormously strong reaction, surely because people are now much more engaged in what kind of food they consume than some decades ago. In some countries, people are even afraid of dead nematodes e.g., Greeks think that dead nematodes will cause strong immunologic reactions and illness if consumed or even cause psychological disturbances in people who see nematodes in fish. It has, however, been shown that dead nematodes are safe, that is, they do not cause illness like anisakiasis.

In Germany, nematodes have been connected to pollution of the North Sea and adjacent waters. This has made matters even worse. There are many people and even some scientists who are—wrongly—of the opinion, that the nematode problem in fish will be solved together with the pollution problem. This is not so. Nematodes are even more numerous in fish in unpolluted waters. Pollution really has nothing to do with the abundance of nematodes. However, abundance of marine mammals has, because they are the final hosts for sealworm and herringworm.

It has also been put forward that consumption of fresh wild fish is dangerous to one's health because of parasites, and therefore that one should rather choose cultured fish as food. On the contrary, in aquaculture it is necessary to use antibiotics and other medicine to keep down all kinds of fish diseases, and it is in fish cultures where many parasites and diseases of fishes proliferate the most. A fresh fish caught wild in clean waters is after all the cleanest and healthiest food available, even if it may contain some nematodes. These can be picked out, and any nematodes left are, after secure processing or cooking of the fish, dead and harmless.

In order to combat misinformation about nematodes in fish proliferated by the media, Danish authorities have undertaken the task of informing the public by publishing a booklet answering some of the most important questions regarding nematodes in fish.

6. CONCLUDING REMARKS

A solution of the nematode problem would be the total absence in fish products of nematodes. This could theoretically be achieved by physical means, biological means, or both. Nematodes have been increasing in fish, and the consumer is much more aware of them now than before. In every fish-processing method, detection and removing of nematodes has now to be taken into consideration. This costs money. Nematodes in fish also make fish markets very vulnerable to damaging propaganda against fish consumption. Propaganda which wrongly connects the nematode problem to pollution of the seas has made matters even worse.

REFERENCES

1. Rohde, K. 1984. Diseases caused by metazoans: Helminths. pp 193–320 In O. Kinne (ed) *Diseases of marine animals*, 4. (Part 1), *Introduction, Pisces*. Biologische Anstalt Helgoland, Hamburg.

2. Möller, H., and K. Anders. 1983. *Diseases and Parasites and Marine Fishes*, Verlag Möller, Kiel.

3. Barnes, R. D. 1974. *Invertebrate Zoology*, 3rd. ed., W. B. Saunders Co., Philadelphia.

4. Croll, N. A., and B. E. Matthews. 1977. *Biology of Nematodes*, Blackie, Glasgow and London.

5. Dagbjartsson, B. 1973. Rannsóknir Varðandi hringormavandamálið, *Tækni tíðindi*, 35:1–11.

6. Ackman, R. G., and R. T. Gjellstad. 1975. Gas-chromatographic resolution of isomeric pentanols and pentanones in the identification of volatile alcohols and ketones in the codworm *Terranova decipiens* (Krabbe, 1978), *Analytical Biochemistry* 67:684.

7. Hauksson, E., and S. I. Björnsson. 1985. Áhrif hringormasýkingar á flakanýtingu þorsks (unpublished research notes, in Icelandic).

8. Platt, N. E. 1975. Infestation of cod (*Gadus morhua* L.) with larvae of codworm (*Terranova decipiens* Krabbe) and herringworm, *Anisakis* sp (Nematoda: Ascaridata), in North Atlantic and Arctic waters, J. Appl. Ecol. 10:437.

9. McClelland, G., R. K. Misra, and D. J. Marcogliese. 1983. Variation in abundance of larval Anisakines, sealworm (*Phocanema decipiens*) and related species in cod and flatfish from the southern Gulf of St. Lawrence (4T) and the Brenton shelf (4Vn), Can. Techn.. Rep. Fish. and Aqua. Sci. 1201:1–51.

10. McClelland, G., R. K. Misra, and D. J. Martcll. 1985. Variations in abundance of larval anisakines, sealworm (*Pseudoterranova decipiens*) and related species, in Eastern Canadian cod and flatfish, Can. Techn. Rep. of Fish. and Aqua. Sci. 1513:1–15.

11. Hauksson, E. 1984. Prevalence and abundance of larvae *P. decipiens* and *Anisakis sp.* (Nematoda Ascaridata) in cod (*G. morhua*) from Icelandic waters, Hafrannsóknir 30:1 (in Icelandic, with an English abstract).

12. Rae, B. B. 1963. The incidence of larvae of *Porrocaecum decipiens* in the flesh of cod, Dept. of Agric. and Fish. for Scotland, Marine Res. Rep. 2:1–28.

13. Rae, B. B. 1972. A review of the cod-worm problem in the North Sea and in western Scottish Waters 1958–1970, Dept. Agric. Fish. for Scotland, Marine Res. Rep. 2:1–24.

14. Wootten, R., and I. F. Waddell, 1977. Studies on the biology of larval nematodes from the musculature of cod and whiting in Scottish waters, J. Cons. Int. Explor. Mer. 37(3):266.

15. Hauksson, E. 1990. Könnun á hringormasýkingu þorsks við Ísland 1985–88. Samanburður við könnunina 1980–81, Hafrannsóknir (in Icelandic, with an English abstract; in preparation).

16. Hauksson, E. 1984. Hringormar í fiski, Fiskvinnslan 2:12 (in Icelandic).

17. Hauck, A. K. 1977. Occurrence and survival of the larval nematode *Anisakis* sp. in the flesh of fresh, frozen, brined, and smoked Pacific herring, *Clupea harengus pallasi*, Journal of Parasitology 63(3):515.

18. Ronald, K. 1960. The effects of physical stimuli on the larval stage of *Terranova decipiens* (Krabbe, 1878) (Nematoda: Anisakidae). I. Temperature, Canadian Journal of Zoology 38:623.

19. Hauksson, E. 1989. Geymsla þorsks í krapa and hringormar, Fiskvinnslan 1:29 (in Icelandic).

20. Valdimarsson, G., Einarsson, H., and, F. J. King. 1985. Detection of parasites in fish muscle by candling technique, J. Assoc. Off. Anal. Chem. 68(3):549.

21. Hauksson, E. 1988. Er munur á hringormafjölda í lifandiblóðguðum and dauðblóðguðum fiski, Fiskvinnslan 1:15 (in Icelandic).

22. Hauksson, E. 1989. Staðsetning hringorma i þorskflökum, Fiskvinnslan 2:14 (in Icelandic).

23. Power, H. E. 1958. The effect of various lighting conditions on the efficiency of "candling" cod fillets for detection of parasites, J. Fish. Res. Bd. Canada, 15 (4):537.

III

Lighting and Vision Techniques

6

Evaluation of Illumination Methods for Machine Vision Applications in the Fish Industry

TIMO PIIRONEN *Technical Research Center of Finland, Oulu, Finland*

1. Introduction
2. Backlighting
3. Polarized Front Lighting Method
4. Specular Dark Field Illumination
5. Omnidirectional Illumination
6. Structured Lighting
7. Non-Camera-Based Methods
8. Summary

   References

1. INTRODUCTION

The prerequisite for any vision application is that the features to be analyzed by the computer can be seen in the image. In spite of the rapid progress in sophisticated image analysis algorithms, the performance of the camera and illumination subsystem can greatly affect the reliability and cost of a machine vision application. A good illumination design can simplify computer processing by enhancing the image contrast and suppressing background information. The optomechanical construction of a camera and illuminator unit is essential to the accuracy of measurements and the tolerance of a system to environmental conditions such as mechanical vibrations and dust common in industrial applications.

The design of image acquisition requires careful analysis of physical properties of objects and materials relevant to the vision task (1). Selection of illumination and viewing geometries, light source, optics and sensor components is based on the results of that analysis (Fig. 1).

The vision applications in the fish industry present a real challenge to illumination design. Fishes are shiny, three-dimensional objects producing strong specular components of reflection. Dark back and light belly portions are a natural camouflage in sea environments but make successful implementation of simple segmentation methods like video signal thresholding difficult (Fig. 2).

In order to get a tentative understanding of the reflectance behavior of the fishes, several practical illumination designs intended for lighting specular objects were tested with fish samples. The experimental results together with measured data of optical properties of different fish species and detailed functional

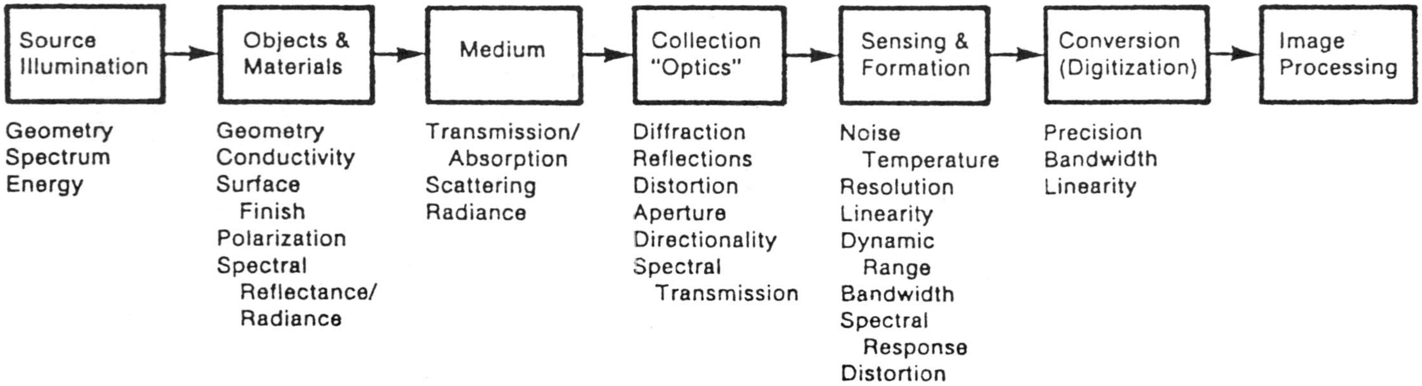

Figure 1 The different phases of image acquisition and related physical properties.

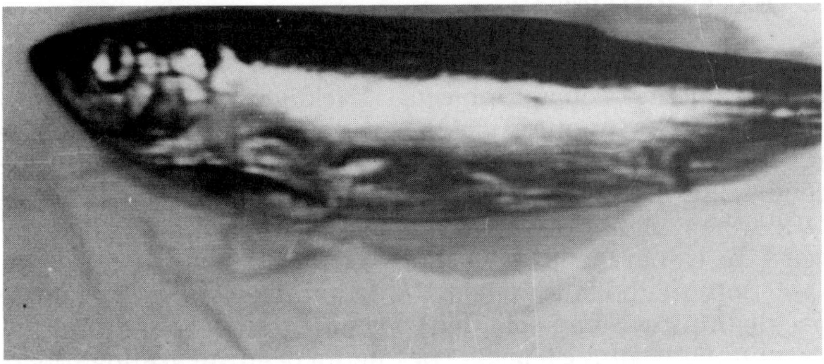

Figure 2 A fish in normal room lighting.

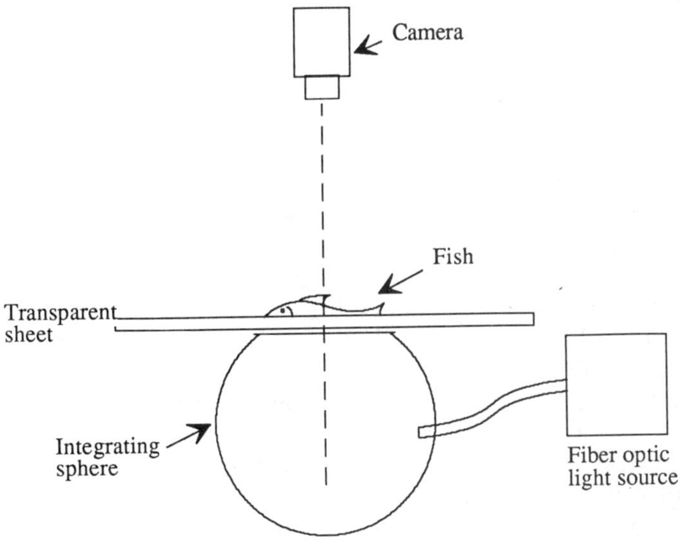

Figure 3 A diffuse backlighting arrangement used in tests.

and accuracy requirements of a particular application can be used for directing further development.

2. BACKLIGHTING

The most obvious illumination method for width and length measurements of opaque objects is backlighting. Two basic variations of the technique exist (2). In the diffuse method a diffusing sheet is placed between the object and the light source. In the condensing method a lens, typically a Fresnel lens, is placed between the light source and the object in order to direct light toward the camera lens.

Both methods were tested with sample fishes. The diffuse backlighting was implemented by an integrating sphere and a fiber optic light source (Fig. 3). The interior of an integrating sphere is coated with a diffuse matt paint. Multiple reflections from the walls of the sphere produce very even and diffuse illumination on the exit port of the sphere (1). For the condensing illumination a fish was placed on the fresnel lens and illuminated behind the lens with fiber optic light source. The lens formed the image of the fiber optic bundle to the entrance pupil of the camera lens (Fig. 4).

In diffuse backlighting (Fig. 5), partly translucent parts of the fish such as the tail and the head have a low contrast in respect to the background. Also, the edge of the belly is somewhat smeared by the specular reflections from the round fish surface.

In condensing illumination the light is directed and specular reflections from the fish are avoided (Fig. 6). Very sharp edges are formed, and the translucent areas of the fish are as dark as the main body. However, the method is very sensitive to dirt or water on the background sheet. Water drops on the fresnel lens act like small lenses, deflecting the light away from the camera lens, and can affect width and length measurements catastrophically (Fig. 7).

Both backlighting methods seem to have drawbacks regarding accurate length and width measurements. However, in spite of uncertainty in exact edge position due to translucency and specular reflections, the diffuse backlighting is more suitable because of its robustness. Designs insensitive to mechanical vibrations are also easier to produce for diffused lighting than for condensed lighting.

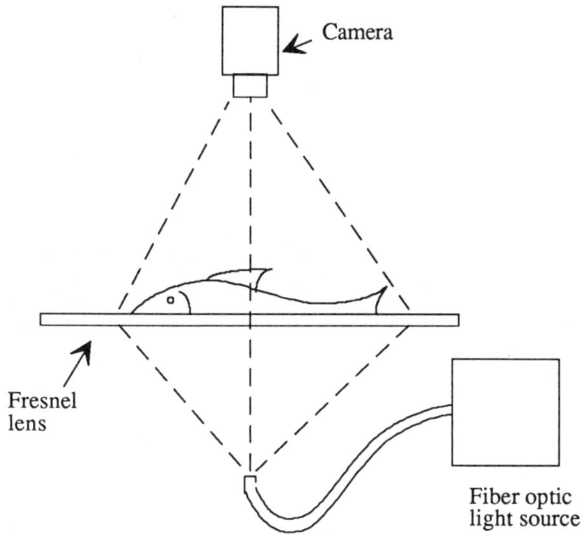

Figure 4 A condensing backlighting arrangement.

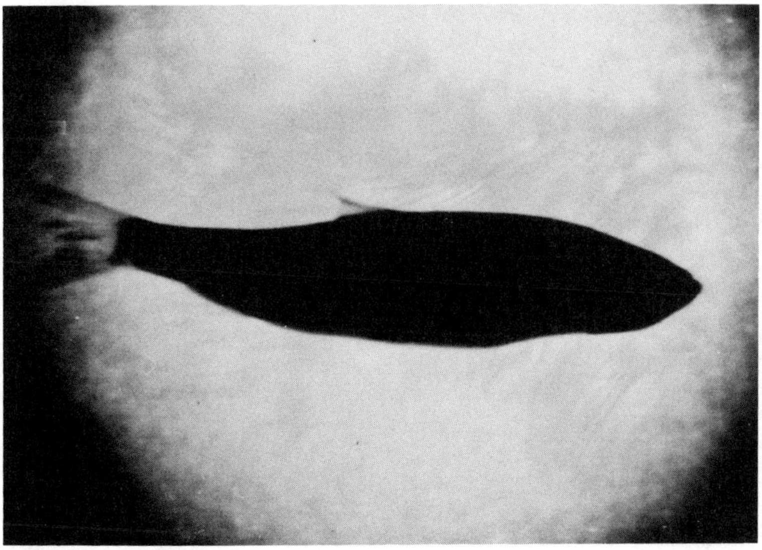

Figure 5 A fish backlighted diffusely.

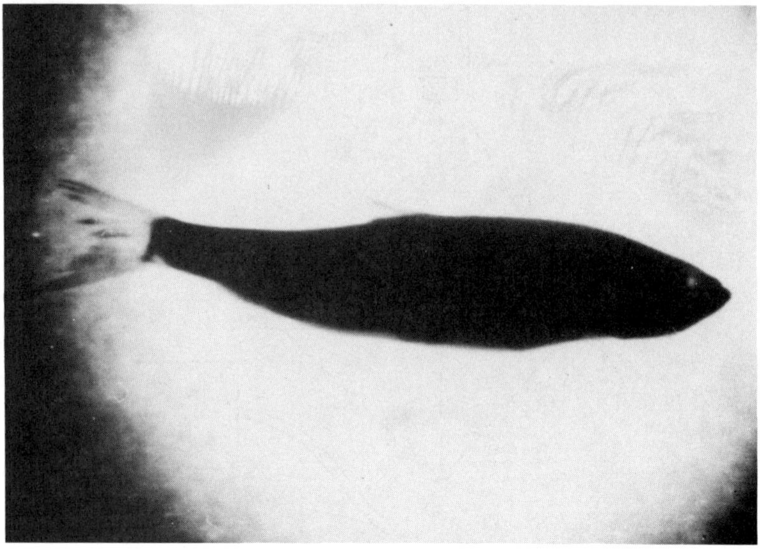

Figure 6 A fish imaged with condensing backlighting.

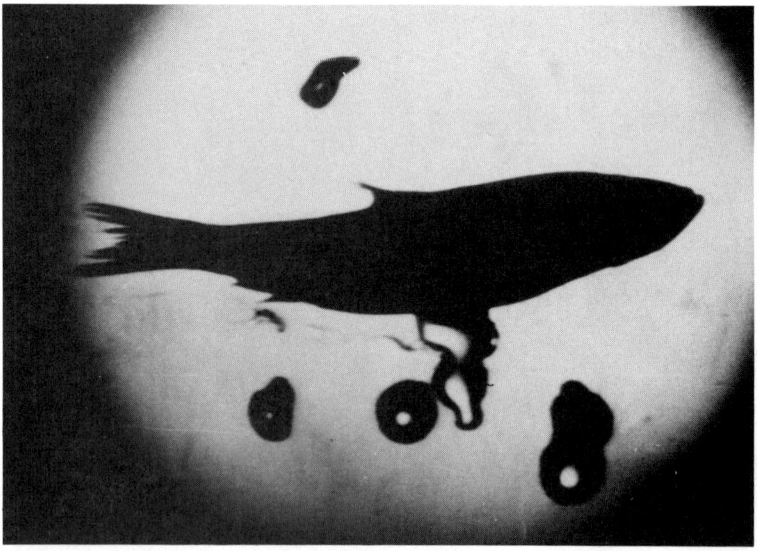

Figure 7 A fish and water drops with condensing backlighting.

The prerequisite for backlighting is that the conveyor belt transporting fishes is at least partly transparent. A line scan camera–based implementation produces a compact design because the illuminated area can be small in the direction of belt movement. For length measurements the speed of the conveyer belt has to be measured or the exposure time of the camera time has to be synchronized to the line speed.

The diffuse backlighting for line scan camera can be implemented by using linear light sources like fluorescent lamps. Because of strong AC-modulation in the light output, fluorescent tubes have to be driven with high frequency power supplies when short exposure times are used. If the brightness of a fluorescent tube is too low, e.g., due to the low transmission of a conveyer belt, they can be replaced by linear low pressure sodium lamps having radiances approximately a decade better than the fluorescent tubes. Sodium lamps also require high frequency power supplies. These are not as commonly available as high frequency power supplies of fluorescent tubes that are used in vehicle illumination (3).

For very high speed or high intensity applications, linear halogen lamps can be used. High power (500–2000 W) linear halogen lamps produce a great amount of heat, which must be considered in the design. AC modulation in light output of halogen lamps is approximately 10%. The DC power for main voltages can be easily produced by full wave rectification of three-phase AC current.

When moving objects are imaged by normal video cameras, the fixed exposure time of a camera can cause problems. The exposure times for CCD video cameras are typically 20 ms or 40 ms when the European video standard is used. If the speed of a conveyer belt is 0.5 m/s, an object moves 20 mm during 40 ms exposure time, resulting in smearing images. There is also a 20 ms difference in exposure timings of the odd and even fields of a standard video frame, adding distortion to the image. These problems are solved by stroboscobic illumination (4). A short, typically 15 μs light pulse from xenon flash lamps during the vertical blanking period of the video signal is used to freeze the motion of an object. Because the xenon flash lamps have very high brightness and good radiation efficiency, they are capable of removing the effect of normal room lighting during the exposure times. If the background illumination is very strong, it can be eliminated by mechanical shutters synchronized to the light pulse triggering. Special cameras with electronic shutters are also available.

3. POLARIZED FRONT LIGHTING METHOD

A well-known method for the elimination of specular reflections is to illuminate an object with linearly polarized light and remove the polarized light component from the reflected light (5). The basic assumption is that specularly reflecting surfaces usually maintain the polarization of the incoming light. Rough diffusely scattering surfaces act as depolarizers, destroying the polarization of the reflected light.

The test arrangement is depicted in Figure 8. Nearly coaxial illumination from above reduces the effect of a incident angle to the polarization (6). Linear sheet polarizers were used. The ratio of transmittance of maximum rejection for a linearly polarized light component to that of maximum transmittance for linearly polarized light was less than 1/200. The results of the experiments are depicted in Figures 9, 10, and 11. In order to make the effects of polarization and water on the fish more apparent, a specularly reflecting stainless steel spoon is used in the figures as a reference object. As can be seen in Figure 10, polarization-based rejection of specular reflection works reasonably well with dry fish surfaces. Unfortunately, the skin of a fish is often wet, and in that case the effect of polarization is very weak (Fig. 11). The explanation for the phenomenon is that the liquid formed by the water and solvents from the skin of a fish depolarizes light. Experiments suggest that the straightforward utilization of polarized light cannot be used for elimination of specular reflections from the fishes. Probably polarization could be used to evaluate the moisture of fish skin if that information is needed.

In spite of strong specular reflections in the image, nearly coaxial illumination is not totally useless for length and width measurements. There is a good contrast between fish edges and background. Provided that the viewing angle of the camera can be kept reasonably low, specularly reflecting areas are located inside the main body of a fish having no effects on the positions of fish edges. If it is possible to affect the reflectivity of a conveyer belt so that the intensity of the background area is between the

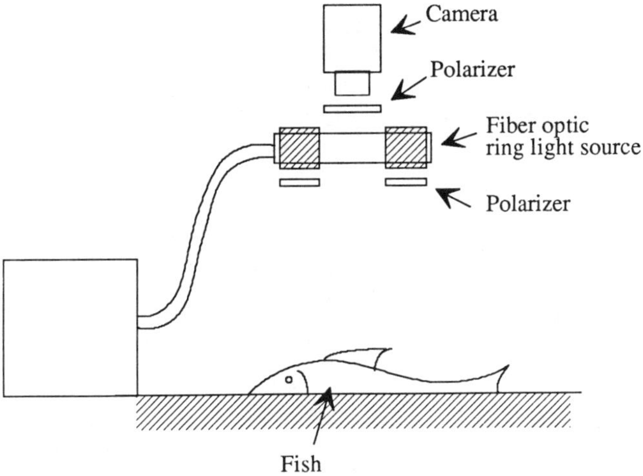

Figure 8 A polarized lighting arrangement.

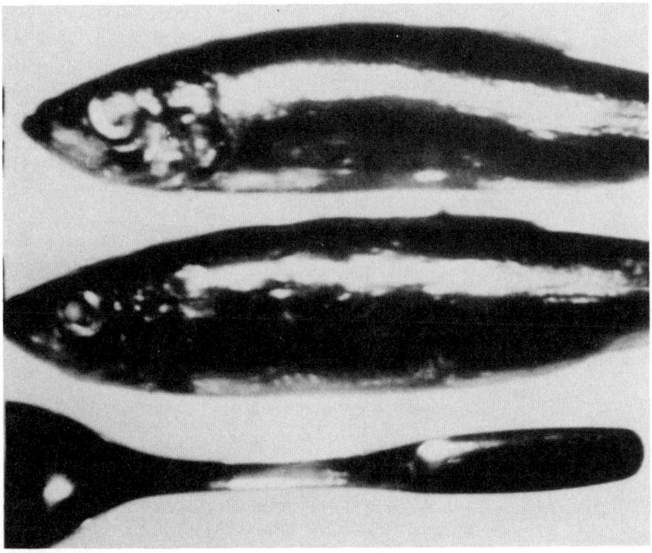

Figure 9 Front lighting with polarized light when the analyzer and the polarizer have the same polarization orientation.

Figure 10 Dry dishes under polarized illumination when the analyzer is turned for maximum rejection of specular reflections.

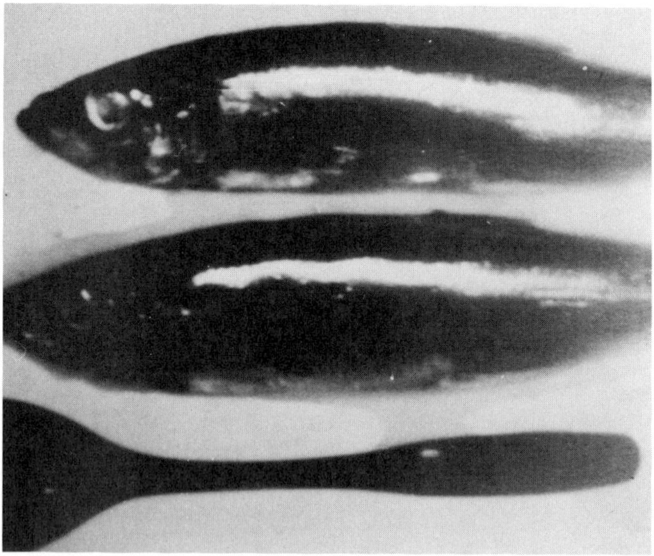

Figure 11 Wet fishes under polarized lighting when the analyzer is turned for maximum rejection of specular reflections.

intensity values of dark and specular areas of a fish, most of the specular areas can be differentiated by simple thresholding. For a transparent conveyor belt, a sheet with appropriate reflectivity under the belt can be used to control the background reflectivity. If a sheet with a bright color is selected and fishes are imaged with a color video camera, it is also possible to differentiate the background by comparing the intensity ratios of different color channels.

4. SPECULAR DARK FIELD ILLUMINATION

In the dark field illumination method, the direction of the incident light and the viewing angle of the camera are selected so that most of the specular reflections are away from the camera. An arrangement depicted in Figure 12 was tested by fish samples. The performance of the method is dependent on the slope variations of the object surface. For round three-dimensional objects like fishes it is difficult to find a suitable nonspecular viewing and lighting direction. The best results were obtained when fishes were oriented with tails toward lighting and camera (Fig. 13). The contrast of the fish belly is quite low, which can cause problems in edge measurements. If the orientation of the fish is changed, specular reflections make accurate dimensional measurements difficult (Fig. 14). Light from other light sources near the inspection area must be blocked out because it can also cause specular reflections in the image.

5. OMNIDIRECTIONAL ILLUMINATION

Specular reflection spikes can be eliminated by evenly directing illumination on the object from the whole hemisphere surrounding it. That can be implemented by integrating sphere based illumination (Fig. 15). On very smooth specular surfaces, such as a fish eye and a stainless steel spoon, the images of a camera viewing port can be seen as darker areas (Fig. 16). If required, it can be eliminated by placing a beam splitter above the camera port to produce coaxial illumination (7). Omnidirectional illumination removes the effect of light direction in the image and greatly re-

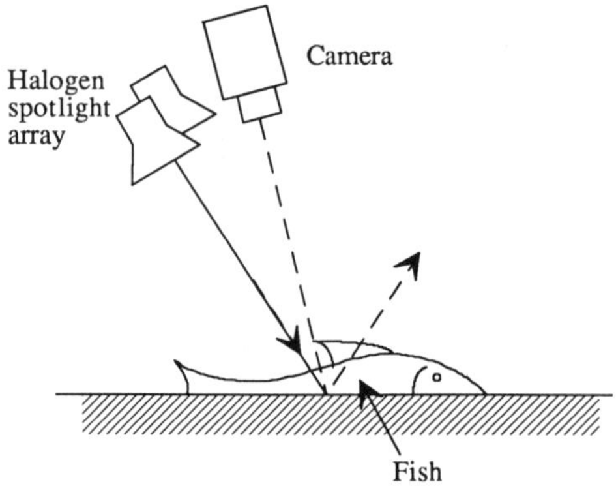

Figure 12 A specular dark field illumination arrangement.

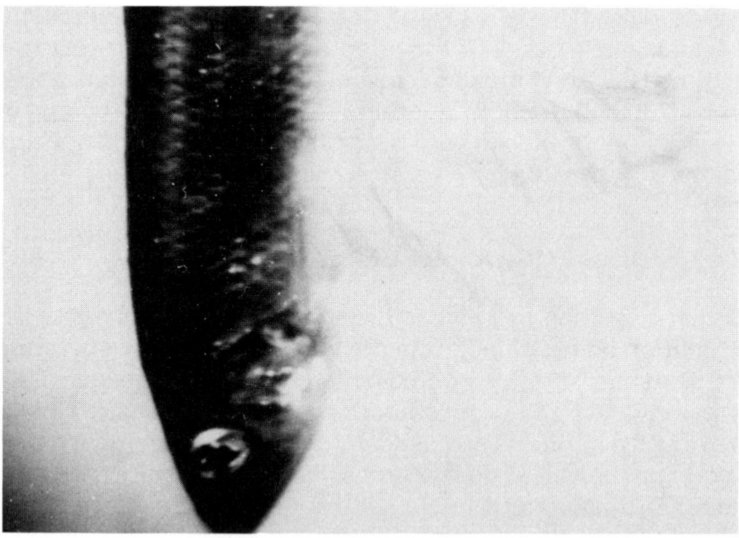

Figure 13 A fish under dark field illumination when light comes from the direction of the tail.

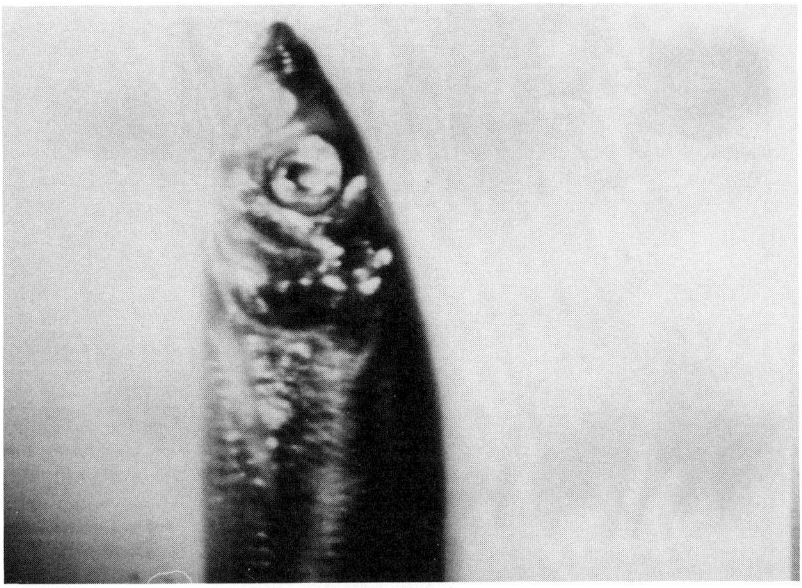

Figure 14 Fish under dark field illumination when light comes from the direction of the head.

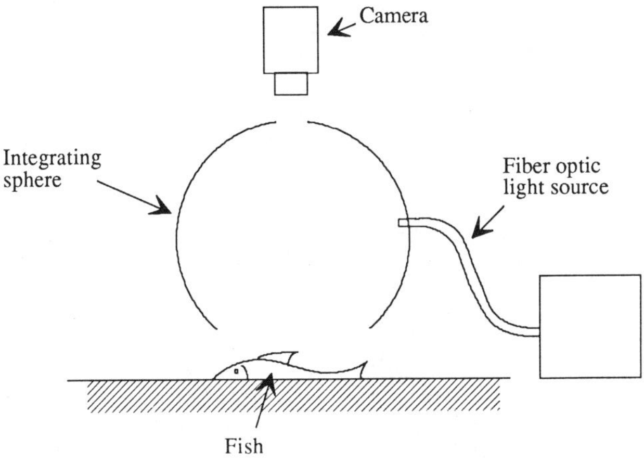

Figure 15 The omnidirectional illumination arrangement.

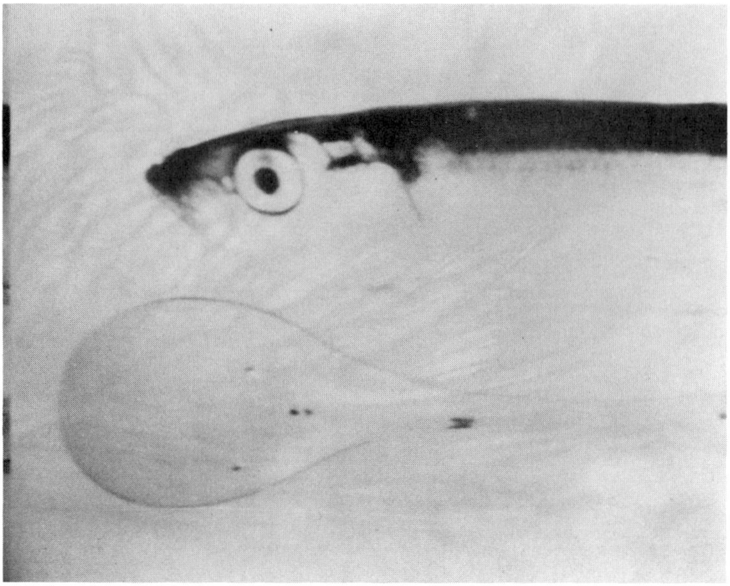

Figure 16 A fish and a stainless steel spoon under omnidirectional illumination.

duces the effect of object shapes. Image contrast is mainly determined by the total reflectivity of objects.

The reflectivity variations of sample fishes were so great that it was difficult to get a good contrast for all the edges with an integrating sphere–based illumination. Spectrally both back and belly areas of sample fishes has no strong color components in the visible spectrum when evaluated by eye. Therefore, spectral optical filtering could not be effectively used to decrease the contrast between the different parts of a fish. However, actual measurements are needed for optimum spectral range selection. If the reflectivity of the background can be chosen between the intensity values of back and belly portions of a fish, edges can be segmented by using two thresholds. Color encoding of the background material can also improve edge measurements.

Stabilizing the intensity variations the omnidirectional illumination is useful for measurements of spectral properties. For

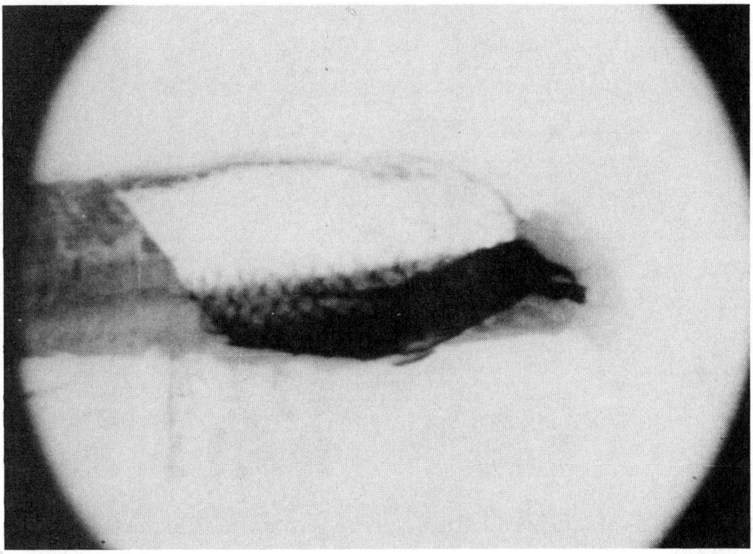

Figure 17 A piece of skin on a fish fillet under omnidirectional illumination.

example, blood spots can be detected by using optical filters or color cameras.

Under integrating sphere–based illumination, the intensity level of the fish flesh was between those of dark and bright portions of fish skin that can be used for detecting skin on fish fillets (Fig. 17).

The diameter of the exit port of the integrating sphere used in experiments was approximately 50% of the sphere diameter. If the dimensions of the sphere are a problem in application environments, similar effects can be obtained with a hemisphere or a cone.

6. STRUCTURED LIGHTING

By projecting a light stripe on an object its three-dimensional shapes can be measured when the angle between camera and light source and the distance to the reference plane are known

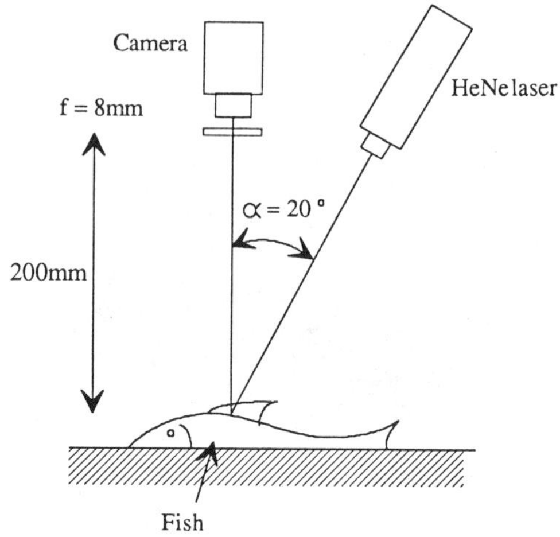

Figure 18 A structured lighting arrangement for the measurement of fish height, width, and length.

(Fig. 18). Using this method the thickness of the fish can be measured at the same time as the length and width. When a monochromatic laser light source is used, the background illumination can be removed by optical filtering (Fig. 19 and 20). The depth accuracy of the experimental arrangement depicted in Fig. 18 was approximately 0.6 mm at the center of the image.

In Figure 20, problems caused by shadowing and multiple reflections related to the triangulation principle are clearly seen. They can cause ambiguity if the volume of the fish is to be measured. However, the width and maximum height information can be obtained by a relatively simple process. If the movement of a conveyer belt in vertical direction can be removed, e.g., by placing a roller under the belt, the knowledge of the height position of the belt can be used to simplify algorithms. With a standard video camera the minimum sample period is 20 ms. When the line speed is 0.5 m/s, a 10 mm ambiguity to the length measurements results. This is similar to ambiguities due to shadowing at the fish ends.

If only the width and length information is needed, the processing can be simplified by replacing an area camera with a line

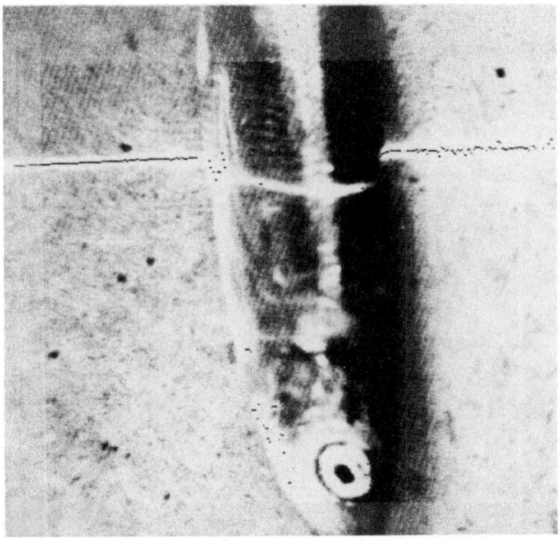

Figure 19 A laser stripe on the fish under room lighting.

scan camera. By aligning the camera so that it sees the light stripe when there are no objects on the belt, all three-dimensional objects are seen black in the image. In the experiments a fish was placed on specularly reflecting material and illuminated with an array of spotlights (Fig. 21). The illumination arrangement was modified from illuminators developed for the inspection of metallic strips in rolling mills (8). A very strong contrast was achieved, and a fish body could be segmented by simple thresholding in spite of water drops on the background material

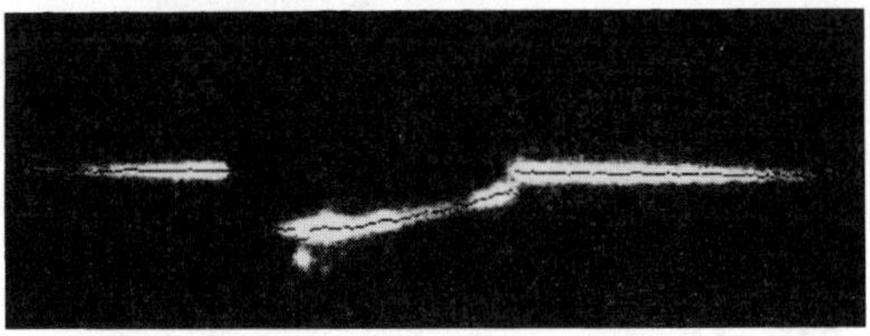

Figure 20 Background illumination removed by an interference filter.

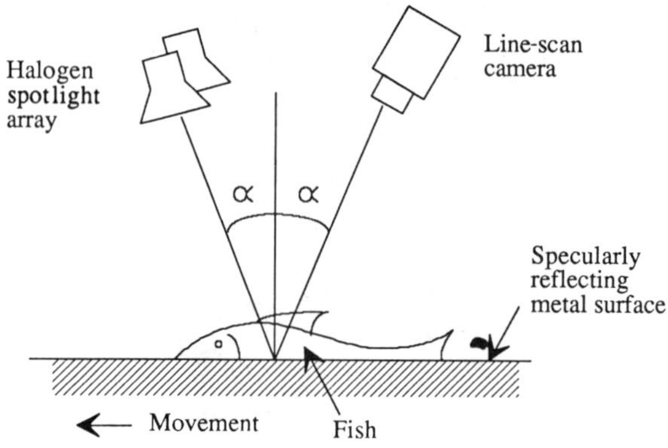

Figure 21 The specular bright field illumination arrangement.

Figure 22 A fish and a stainless steel spoon under specular bright field illumination.

(Fig. 22). For diffuse belt materials a light stripe pattern must be used.

7. NON-CAMERA-BASED METHODS

Width and height information can be obtained without a camera by using a long detector array and projecting a shadow of an object on the array with a collimated light (9). A modified version of the method uses arrays of detectors and light emitting diodes on both sides of an object (Fig. 23) (10). The LEDs and detectors are scanned by activating one light source and three detector pairs at a time. Light rays from light sources to detectors form a net in the measurement area, and a large depth of field is obtained. The geometry of detectors and light emitters can be arranged so that the dimensions of a object can be determined simply by calculating

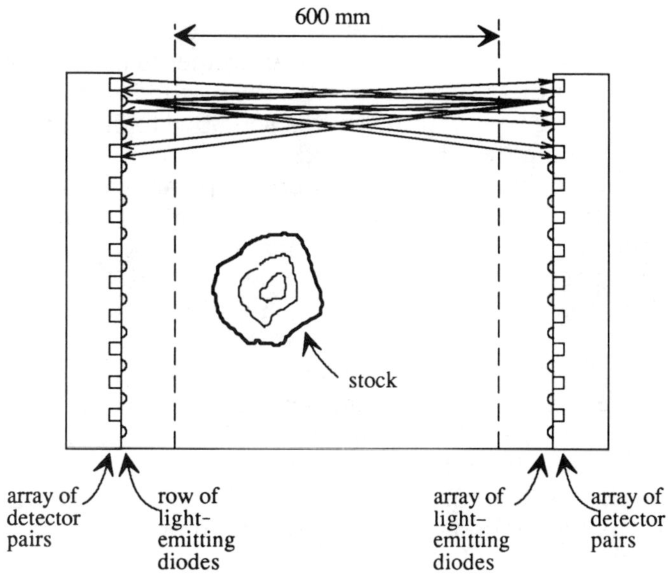

Figure 23 A long depth of field dimension measurement with LED and detector pairs.

how many rays are not detected by corresponding detectors. For vision applications the system can be considered as a binary camera producing a silhouette of the object when it moves through the measurement area.

This principle has been applied to measure diameters of stems in forest harvesters used in very harsh outdoor conditions. The developed measuring device is insensitive to mechanical vibrations, snow, oil, and dirt encountered frequently in those environments. The measurement accuracy of 1.5 mm was obtained with a 6 mm spacing of LEDs and 40 mm spacing of detector pairs. The matrice of 200 LEDs and 60 detectors can be scanned in 2 ms. In addition to the robustness and large depth of field, the advantage of the method is the compact structure, because the minimum working distance required is much shorter than with camera-based solutions. Possible applications of the method in the fish industry are the measurements of fish width and height profiles on trawlers or in other demanding environments.

8. SUMMARY

The transparency of the fish and the specular reflections can cause inaccuracy in dimensional measurements of fishes when diffuse backlighting is used. However, the method is more robust than condensing backlighting that is sensitive to dirt and water drops on a transparent background.

Elimination of specular reflections by polarized light is difficult due to the depolarizing properties of a wet fish skin. Nearly coaxial illumination from above produces a good contrast between fish edges and background. Specularly reflecting areas can be kept inside the fish body by using a low view angle of the camera.

Most of the specular reflections can be removed from the image if the angle of illumination and the viewing angle of the camera are inclined to the tail of the fish. However, the method is sensitive to the orientation of the fish.

The effect of light direction and fish shape can be greatly reduced by using an integrating sphere–based omnidirectional illuminator. The method is best suited to monitor the spectral and total reflectance properties of fish and fillet surfaces.

Light stripe–based triangulation can be used to measure the height, width, and length of a fish at the same time. The method can be simplified in the width and length measurements by using line scan camera instead of a standard video camera.

The width, height, and length measurement solution based on the use of light emitting diode and detector arrays is a viable alternative to the camera-based imaging in harsh environmental conditions.

REFERENCES

1. Grum, F. and R. J. Becherer, 1979. *Optical Radiation Measurements*, Vol. 1 *Radiometry*, Academic Press.
2. Batchelor, B. G., D. A. Hill, and D. C. Hodgson, 1985. *Automated Visual Inspection*, IFS Publications Ltd.
3. Kaufman, J. E., and J. F. Christensen, 1984. *IES Lighting Handbook 1984*, reference volume. Illuminating Engineering Society of North America.
4. Novini, A., 1987. Fundamentals of strobe lighting for machine vision. Vision 87 Conference, June 8–11, 1987 Detroit, Mich., pp. 4/13–4/25.
5. Merch, S. H., 1984. Polarized lighting for machine vision applications. RI/SME Third Annual Applied Machine Vision Conference, February 1984, pp. 4/40–4/54.
6. Born M., and E. Wolf, 1959. *Principles of Optics*, Pergamon Press.
7. Silvén, O., I. Virtanen, and T. Piironen, 1985. Experimental system for the inspection of printed wiring boards. Intelligent Robots and Computer Vision, September 16–20, 1985, Cambridge, Mass., SPIE Vol. 579, pp. 554–559.
8. Piironen T., O. Silvén, T. Laitinen, E. Strömmer, and M. Pietikäinen, 1988. An automated system for metal strip inspection. IAPR Workshop on CV, Special Hardware and Industrial applications, October 12–14, 1988, Tokyo, pp. 397–400.

9. Mihalik, A. S., and J. Vala, 1988. Shadow vision, SPIE, Vol. 956, Piece Recognition and image processing, pp. 56–61.

10. Rieppo, K., and H. Jokinen. Electronic device for measurement of stem diameter, Metsateho Review 11 (in Finnish).

7

Lighting Setup in the Automatic Detection of Ventral Skin and Blood Spots in Cod Fish Fillets

KEPA BENGOETXEA *Dow Chemical Company, Tarragona, Spain*

◆◆◆

1. Background
   1.1 Purposes of Lighting
2. Detection of Black Skin and Blood Spots in Cod Fish Fillets
   2.1 Spatial Distribution of the Light
   2.2 Temporal Distribution of the Light
   2.3 Reflectivity
   2.4 Spectrum of the Radiation Illuminating the Object
   2.5 Spectral Responses of the Various Optical Components

3. Conclusion
   Appendix I
   Appendix II
   Appendix III
   Appendix IV
   References

♦♦♦

1. BACKGROUND

In any visual inspection problem the quality of the lighting is paramount: Too often key features are obscured by glare, or the intensity of light reaching the detector is insufficient. The best method of illumination is not always obvious.

There are a few general principles in this subject which a good illumination system designer will usually need to consider (1):

- The output light level from the lamps.
- The temporal stability of this light level.
- The spectrum of the radiation illuminating the object under examination.
- The spectral responses of the various optical components and their combination (optics and filters).
- The reflectivity, translucence, and absorption of the object being examined as a function of wavelength (i.e., it color). The possibility of fluorescence should also be considered.
- The spatial distribution of the light, i.e., should it be steady, flashing periodically (stroboscopic), impulsive (on detection of an object for inspection)?
- The polarization of the light.
- The operating environment. In a dusty atmosphere, certain optical components are difficult to keep clean.
- The magnification of the optical system and the field of depth.
- Whether an area-scan or linescan camera is to be used.
- How the object is to be moved past the camera, i.e., is it to be moved continuously, indexed, rotated?

- Ambient light, especially sunlight and light from 50/60 Hz fluorescent lamps.
- Radiation damage to the object being examined, especially by infrared or ultraviolet.

The lighting design should also insure that extraneous lights around the machine are not going to interference with its operation (e.e., daylight) (2).

When designing the optical subsystem, one must try to reduce the demands on the image processor to a trivial level by giving to it the best possible images to analyze. It is almost always cheaper to improve the lighting than the image processing (1).

1.1 Purposes of Lighting

The purposes of lighting are to:

- Improve feature discrimination. The images should provide the vision system with the highest possible contrast between the features and their backgrounds for subsequent analysis (2).
- Reduce processing time. The processing algorithms become simpler by improving the contrast between features and their backgrounds.
- Reduce processing hardware requirements. Often complicated and expensive hardware and software are used to solve a vision problem where some attention to the lighting and optics would have solved the problem (2).

One system to determine the contrast between features will be shown in this chapter. For that purpose, the concept of confusion matrix will be introduced. Assume N samples classified as w1 samples with defect 1, w2 samples with defect 2, and B good samples. The confusion matrix would be as shown in Table 1, where a is the number of samples with the true defect 1 detected by the system as defect 1, b is the number of samples with the true defect 1 considered as defect 2 ones by the system, and c is the number of samples with the true defect 1 considered as good ones by the system (a + b + c = w1); d is the number of samples

Table 1 Confusion Matrix

	TRUTH		
	Defect 1	Defect 2	Good
	a	d	g
SYSTEM	b	e	h
	c	f	i
	w^1	w^2	B

with the true defect 2 considered as defect 1 ones by the system, e is the number of samples with the true defect 2 detected by the system as defect 2, and f is the number of samples with true defect 2 considered as good ones by the system (d + e + f = w2); g is the number of true good samples considered as defect 1 ones by the system, h is the number of true good samples considered as defect 2 ones by the system, and i is the number of true good samples detected by the system as good samples (see Table 1). It is very clear that the higher valued the diagonal divided by the number of samples is, the better will be the detection because the values outside of the diagonal mean wrong detections.

To quantify the detection ability of the system, the correct detection probability and the error probability are as follows:

$$\text{Correct detection probability} = \frac{1}{3} * \left(\frac{a}{w_1} + \frac{e}{w_2} + \frac{i}{B} \right)$$

$$\text{Error probability} = 1 - \text{Correct detection probability}$$

For more defects, the procedure would be the same.

2. DETECTION OF BLACK SKIN AND BLOOD SPOTS IN COD FISH FILLETS

2.1 Spatial Distribution of the Light

Background

Depending on the nature of the features to be inspected,the designer must first choose one of the following techniques in order to solve the spatial distribution problem. These are:

- Front lighting
- Backlighting
- Structured lighting

In front lighting the camera and the light are on the same side of the object (Fig. 1). This technique will be used whenever a surface feature or textures needs to be inspected (3).

Backlighting (Fig. 2) usually provides the vision system with the most contrast and can simplify the algorithm task. However, backlighting does not give information about the surface to be inspected.

In structured lighting, a sheet or dot array of light is directed to the subject at an angle other than the viewing angle (Fig. 3). The intersection of that flat sheet with the subject will follow the contour surface (4). Two applications are shown by Novini (2):

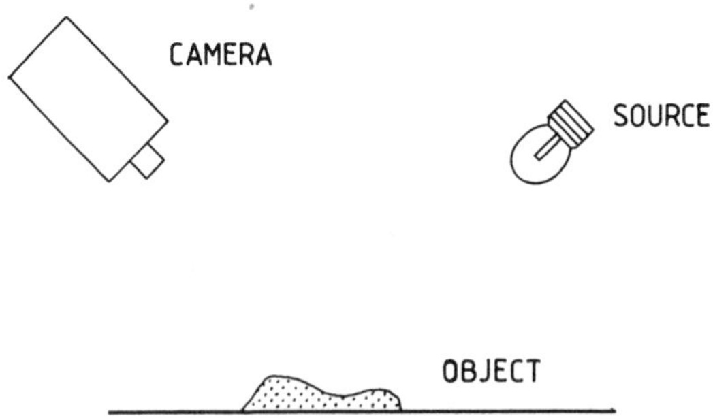

Figure 1 Front lighting.

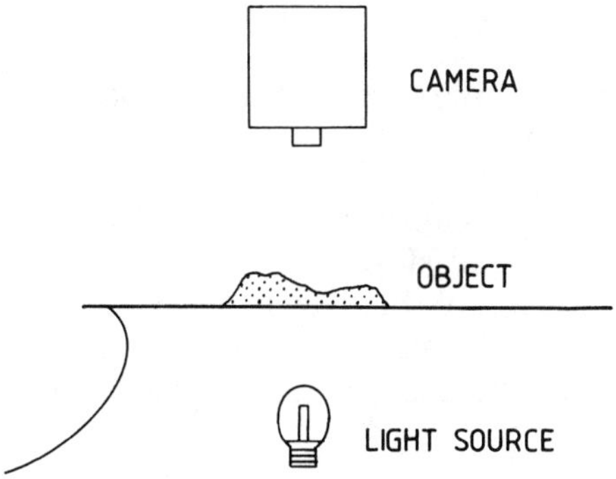

Figure 2 Backlighting.

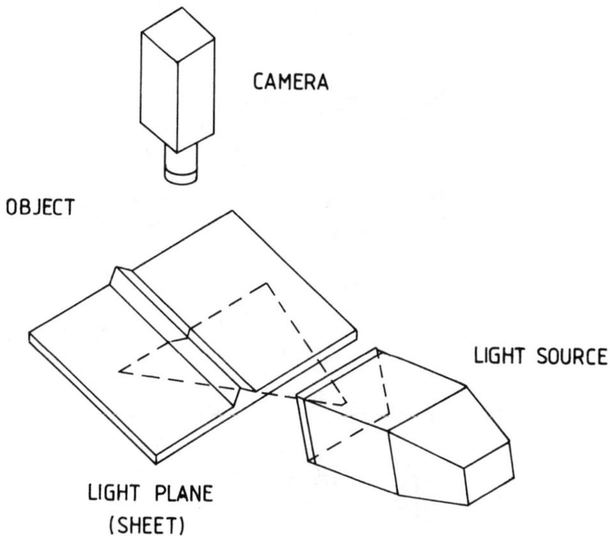

Figure 3 Structured lighting.

- Encode depth information into the two-dimensional scene
- Illuminate some areas more than others by masking the illumination.

The second application would be used in the detection of blood spots and ventral skin providing more accurate information about the location of the defects than using front lighting for trimming purposes. Because the setting is quite expensive and the purpose of this paper is the detection and sorting, front lighting will be considered.

Detection Methods

There are four main problems using front lighting in the detection of blood spots and ventral skin. These are:

- Glinting, which is a specular reflection because of the wet surface of the fish.
- Shadows, give the appearance that the fish has some impurities.
- Unsharp edges between the different features and the flesh.
- Unsharp edges between the flesh and the transport belt.

Batchelor (1) recommended five methods to reduce those problems:

Method 1 Objective: To provide uniform, omni-directional illumination to eliminate shadows (Fig. 4).

Equipment: Circular illumination source (e.g., circular fluorescent tube, driven by a dc or high-frequency supply; fiber-optic device; circular array of small bulbs). Hemispherical diffuser (painted matte white or stipple inside) with viewing aperture at the center.

Typical applications: Viewing complex parts, fibrous parts, etc. The hemispherical diffuser is not reliable for working together with the transport belt in the filleting plant. One alternative to this method is shown in Figure 5.

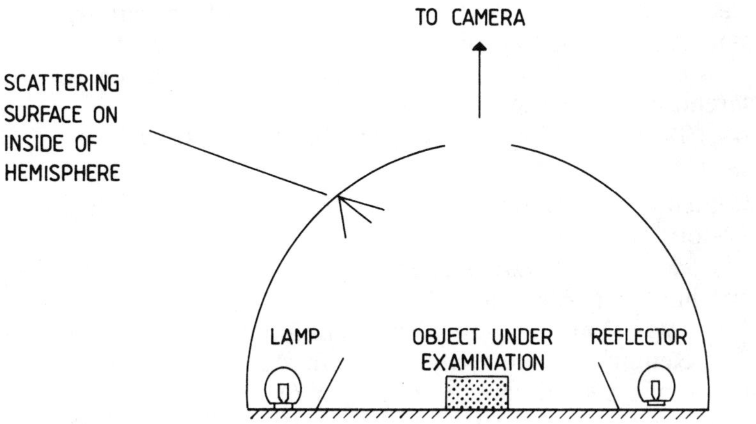

Figure 4 Method 1 (objective).

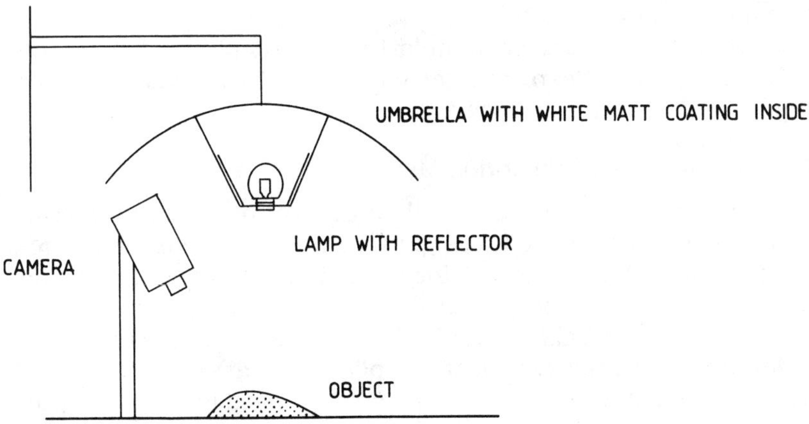

Figure 5 Alternative to Method 1 (objective)—Method 1 (typical applications).

Method 2 Objective: To provide diffuse illumination without causing glinting (Fig. 6).

Equipment: Spot lamps or fiber-optic device. Diffusing screen with viewing aperture.

Typical application: Viewing bright metal surfaces.

Method 3 Objective: To view a flat brightly reflecting surface without glinting.

Equipment: Two or more flood lamps (Fig. 7) or a fiber-optic ring light (Fig. 8).

Typical applications: Examining polished metal surfaces.

Remarks: Ensure that the front face and lens of the camera are in the dark, otherwise a reflection of the camera will be visible.

Method 4 Objective: To reduce glinting (specular reflection) when viewing brightly reflecting materials (e.g., metal, plastics, ceramics), glossy paint, wet and oily surfaces (Fig. 9).

Equipment: Two linearly polarizing filters.

Typical applications: Reading printing on parts of complex and/or unpredictable geometry. In nonpolarized light, printing on such components is difficult to see, because the glinting disturbs the camera's response to the darker parts of the image.

Method 5 Objective: To reduce glinting when viewing brightly reflecting surfaces which also require the provision of shadow-free illumination (Fig. 10).

Equipment: Circularly polarizing filter, plus a collimated light source and beam splitter (pellicle or half-silvered mirror type, not double-prism type).

Evaluation of the Illumination Methods

The criteria for the selection will be the technical requirements (such as shadow free, reduced glinting, and sharp edges), reducing parasitic effects, maximizing the reliability in the processing line, maintenance, and price.

According to Table 2, method 5 is the best from the technical point of view but it is the most complicated to assembly properly and is the most expensive. So, the method used for future study

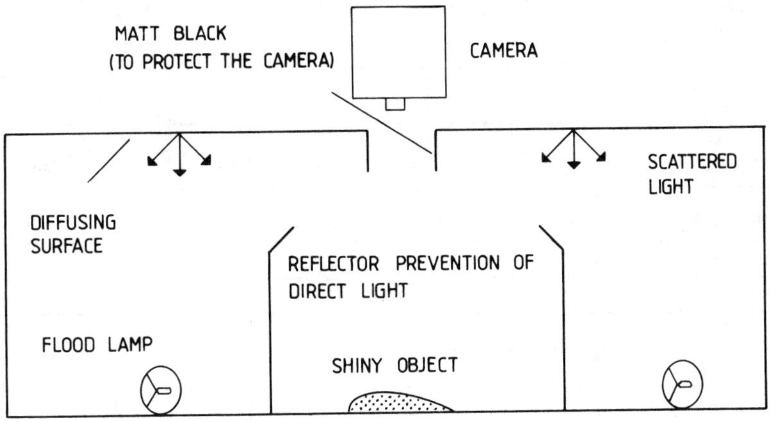

Figure 6 Method 2.

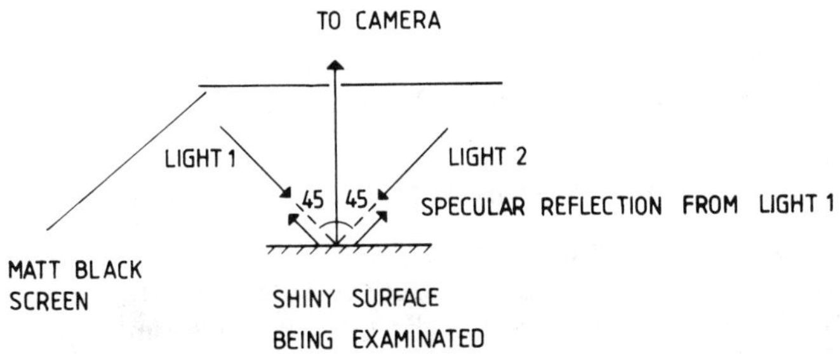

Figure 7 Two-flood lamp solution.

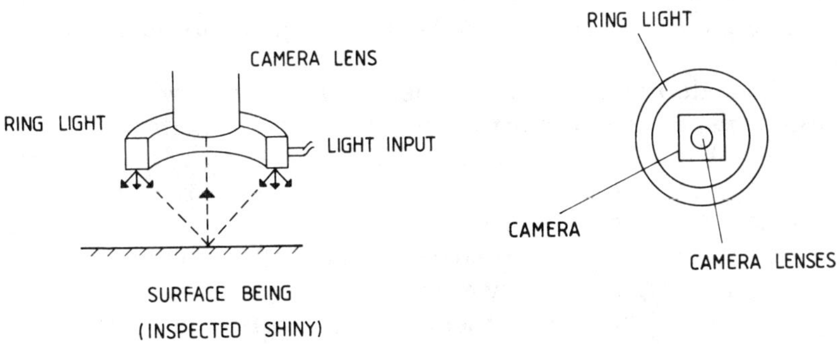

Figure 8 Ring light solution.

Table 2 Evaluation of the Spatial Distribution for Various Illumination Methods

Criteria \ Method	1		2	3		4	5
	Fig. 3	Fig. 4	Fig. 5	Fig. 6	Fig. 7	Fig. 8	Fig. 9
Shadowfree	++	+	+	−	−	+	++
No glinting	+	+	++	+	+	++	++
Sharp edges	+	+	+	+	+	+	++
Parasitic effects	++	+	−	+	+	−	−
Reliability	−	++	++	−	−	−	−
Maintenance	−	+	+	+	−	−	−
Price	−	+	−	−	−	−−	−−

+ = good, − = bad

is the method 1b (see Fig. 5), which is good from the technical point of view, as well as simple and cheap.

2.2 Temporal Distribution of the Light

The fillets are moving continuously in a transport band, so strobe light or an electronic shutter is needed to reduce the effect of image "blurr" that occurs while light photons are being accumulated by the vision sensor during its finite scan period as the fillet moves through its field of view (3) if an area-scan camera is used. This problem is less important with a linear scan camera, but the camera has to be synchronized with the speed of the conveyor belt.

Considering area-scan cameras, there are few options to the designer in order to solve the image "blurr" problem (3). These options include shutters and strobe lights but the idea is the same—by subjecting the detector to a short pulse of light, the effect of image blurr is reduced.

Shutters are used to turn electromagnetic radiation on and off. They accomplish this by mechanically removing and imposing a completely opaque barrier across the optical path. Two basic forms of shutter are normally used. Between-the-lens shut-

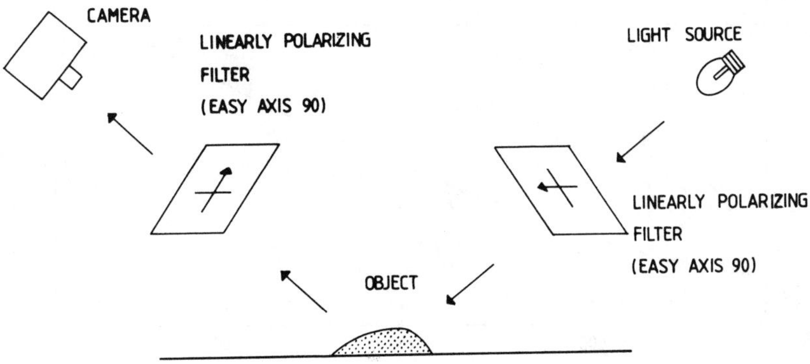

Figure 9 Method 4.

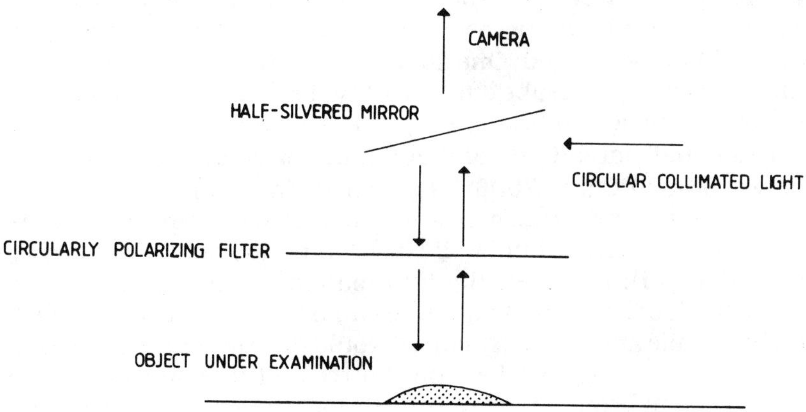

Figure 10 Method 5.

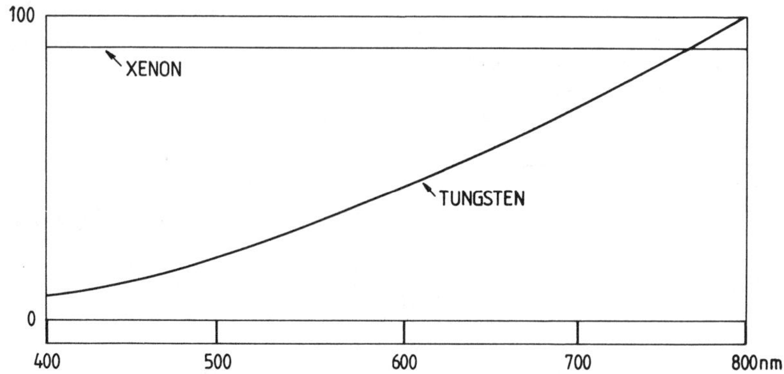

Figure 11 Energy spectrum of two light strobe sources.

ters open and close a symmetric aperture (usually consisting of five thin metal blades) for a predetermined length of time. They were originally designed for specific camera applications. The second common form of shutter is a focal-plane shutter which transports a variable width slit across the focal plane at a fixed (or variable) speed (5). With shutters a very powerful lamp should be used. The shutter speed settings available in the market are 4, 2, 1, 1/2, 1/4, 1/8, 1/15, 1/30, 1/60, and 1/125 sec (typical supplier: MellesGriot).

Any light source that is cycled on and off can be considered a strobe light. However, in order to obtain very short light pulse durations, typically in microseconds, only a certain types of light sources can be utilized. One such source is a gas discharge tube (3). The xenon flash tube emits light typically from 200 nm in the ultraviolet region to 2000 nm in the near infrared region, with peaks usually at 480 nm and 800 nm. The color temperature is between 6000°C and 7000°C (similar to daylight) (3).

The wide spectrum can be very useful in certain machine vision applications, but not if the discrimination between colors is used (Fig. 11). However, the information is needed for sorting purposes, so the xenon flash tube can be used to freeze the fillet motion while another light source would discriminate colors, but of course the image will be a bit "blurred" at that stage.

If the information were needed for trimming purposes, an accurate location of the defects is paramount, so a shutter would

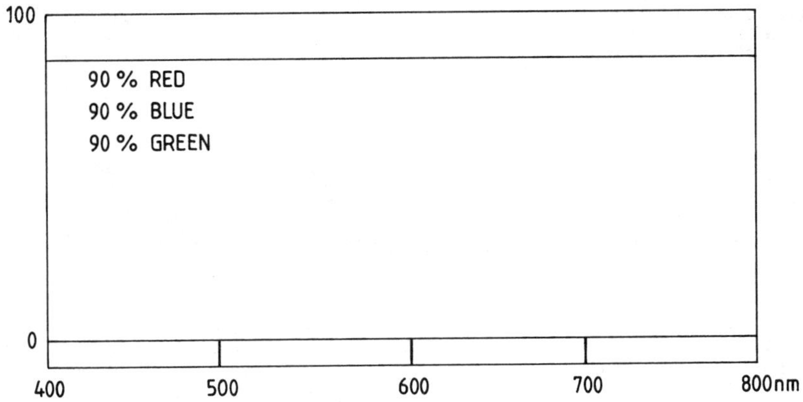

Figure 12 Relative reflectivity spectrum of fish flesh.

2.3 Reflectivity

The relative spectral reflectivity of the features of the fillets was used to discriminate between them.

The features of the fillets are the flesh of white color, the ventral skin of black or greyish color, and the blood spots of brownish or reddish color. According to the color mixing ratios in the Color Video Monitor VR290 of Digital Equipment Corporation and considering the three pure colors from 0% to 100%, the relative spectral reflectivity of the flesh, ventral skin, and blood spots are shown in the Figures 12, 13, and 14, respectively.

2.4 Spectrum of the Radiation Illuminating the Object

Considering the reflectivity requirement (Fig. 15), the lamp should have no components from the yellow to the red part of the spectrum. The problem is where the discrimination point should be set. The best point, where blood spots are enhanced and the contrast between the ventral skin and the flesh is kept, is the green part of the spectrum (line A, Fig. 15). There are no such lamps on the market, so filtering should be used.

Considering the three types of lamps—incandescent (Fig.

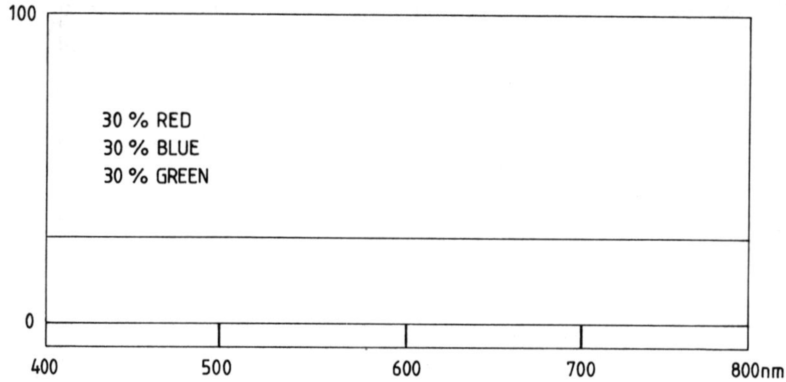

Figure 13 Relative reflectivity spectrum of yellowish flesh (discoloration).

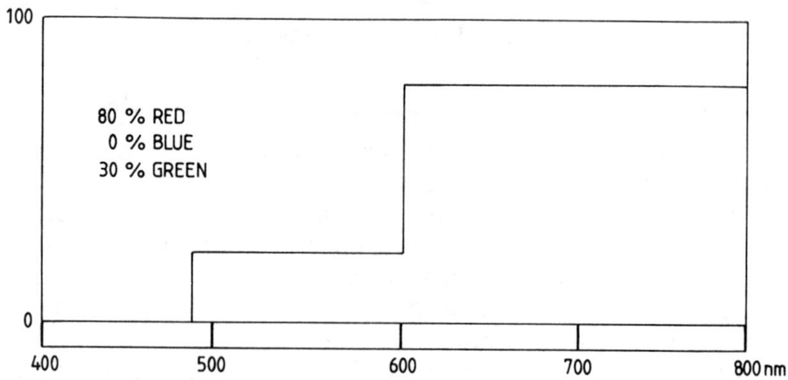

Figure 14 Relative reflectivity spectrum of blood spots.

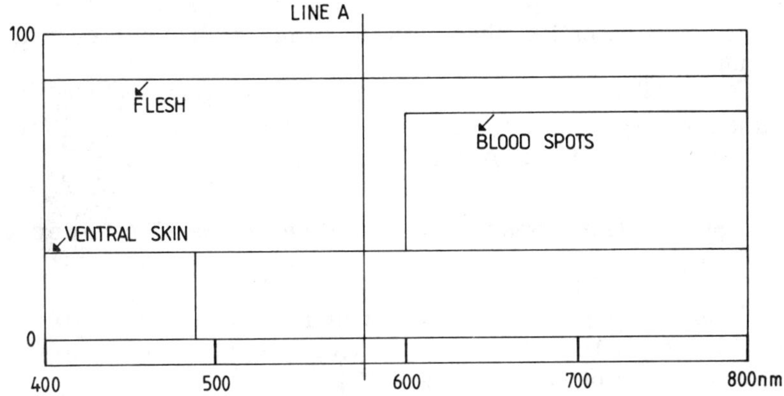

Figure 15 Discrimination between impurities and flesh.

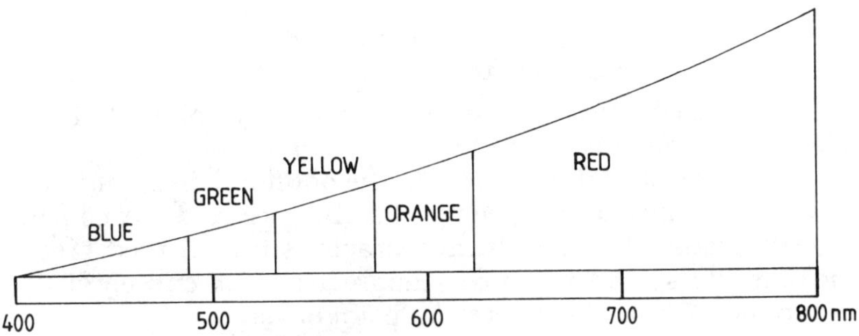

Figure 16 Standard OSRAM incandescent lamp.

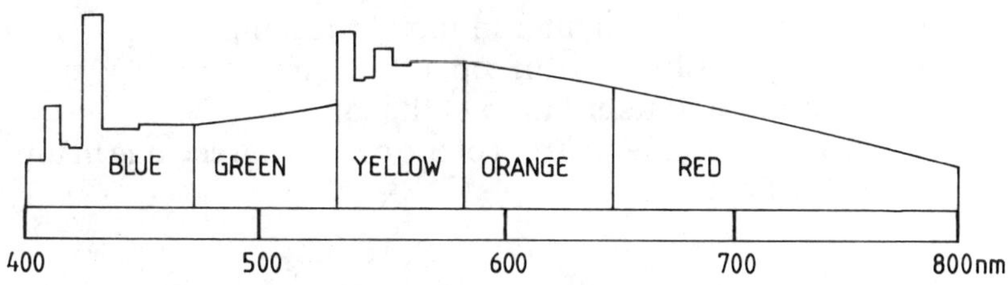

Figure 17 Universal OSRAM fluorescent lamp.

16) fluorescent (Fig. 17), and discharge (Fig. 18)—the last were chosen because of their discrete spectrum, which means easy filtering.

The Osram HQL-R de Luxe 125 W is chosen (Fig. 19), because the lamp should have reflecting coating to avoid the direct light on the fish fillet to reduce glinting (method 1b).

2.5 Spectral Responses of the Various Optical Components

Spectral Response of the Lenses

The lenses have to be chosen according to the minimum absorbance; they should not reduce the electromagnetic radiation in the visible part of the spectrum.

From the catalogue of Melles Griot (5) showing the internal transmittance of three Schott Glasses in the visible part of the spectrum, adjacent curves are obtained (Fig. 20). These curves show that the BK7 or the SF11 lenses are the best for our purposes.

Spectral Response of the Filters

The filters used in these experiments are pieces of hard plastic color plates (red, green, and blue).

As it was shown in Figure 15, the optimum discrimination would be obtained in the green part of the spectrum. So, considering the transmitting spectrum characteristics of the red, green, and blue filters (see Appendix 1), the green one fills entirely all the requirements from the theoretical point of view.

From the practical point of view, different samples of white fish were tested with the various settings of those three filters:

- Downgreen. Green filter in front of the camera
- Downblue. Blue filter in front of the camera
- Downred. Red filter in front of the camera
- Topgreen. Green filter in top of the lamp
- Topblue. Blue filter in top of the lamp
- Topred. Red filter in top of the lamp
- BlGr. Blue filter in front of the camera, green one in top of the lamp

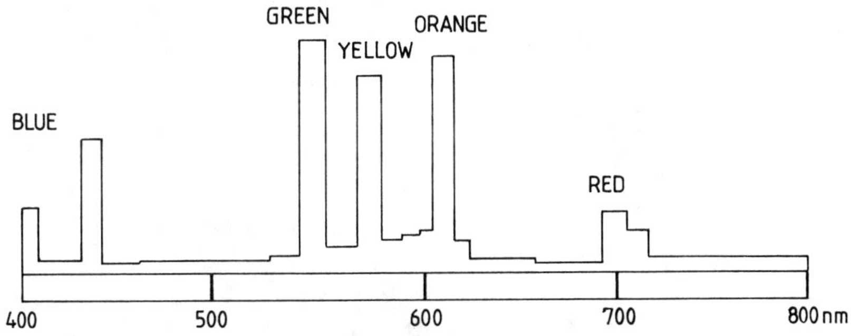

Figure 18 HQL-R Luxe OSRAM discharge lamp.

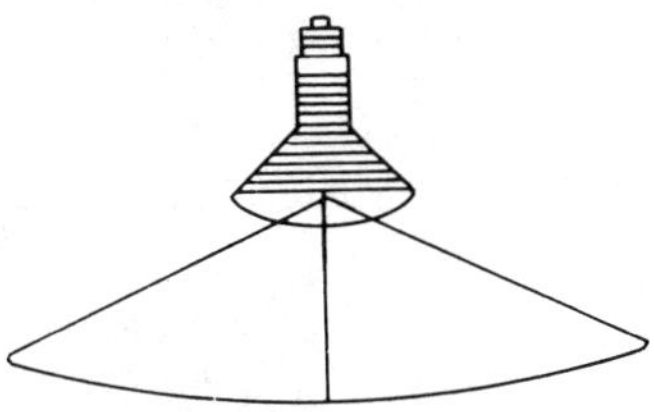

Figure 19 Spread-out angle of the lamp.

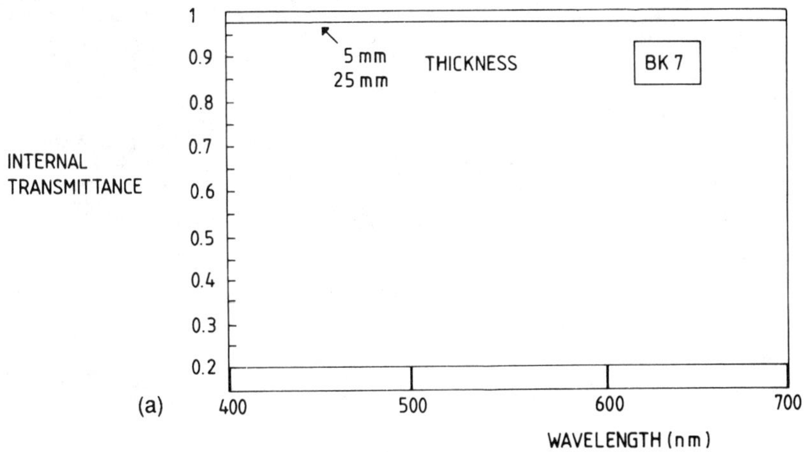

Figure 20 Internal transmittance of three Schott glasses.

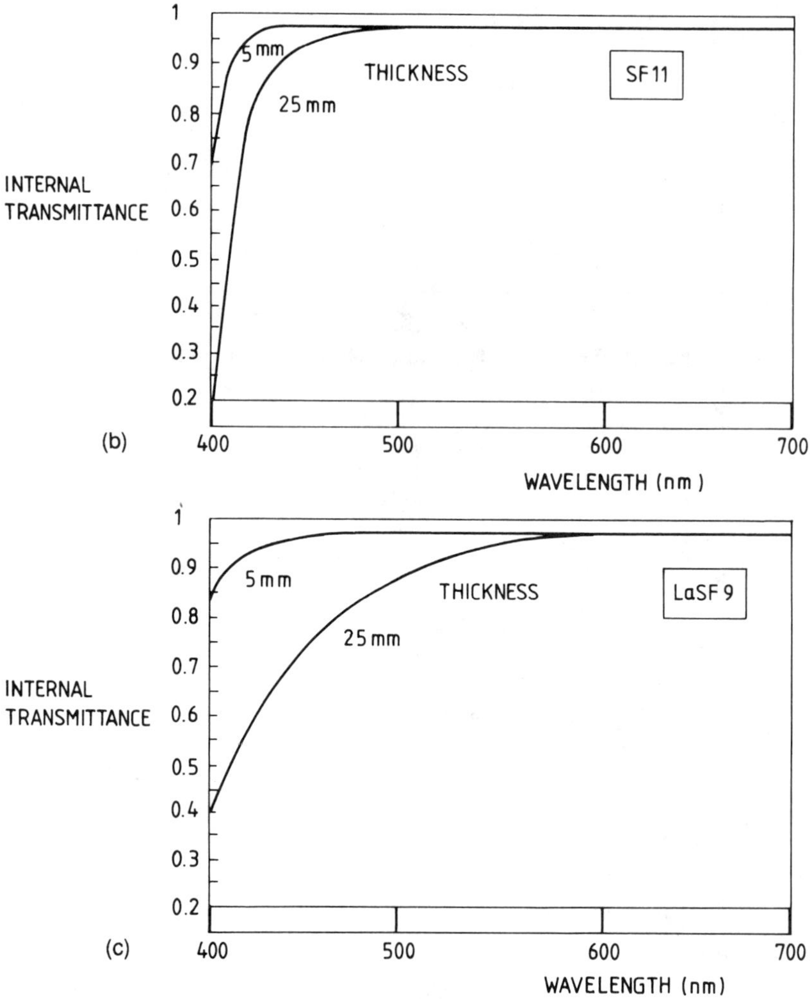

Figure 20 (Continued)

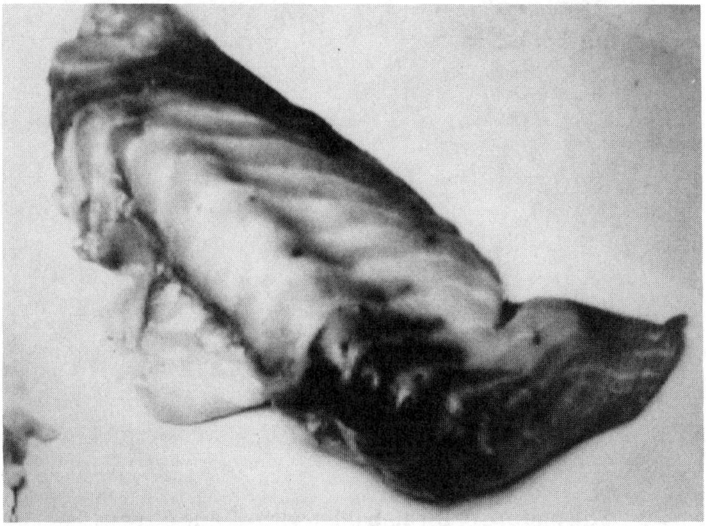

Figure 21 The real picture of the fish fillet.

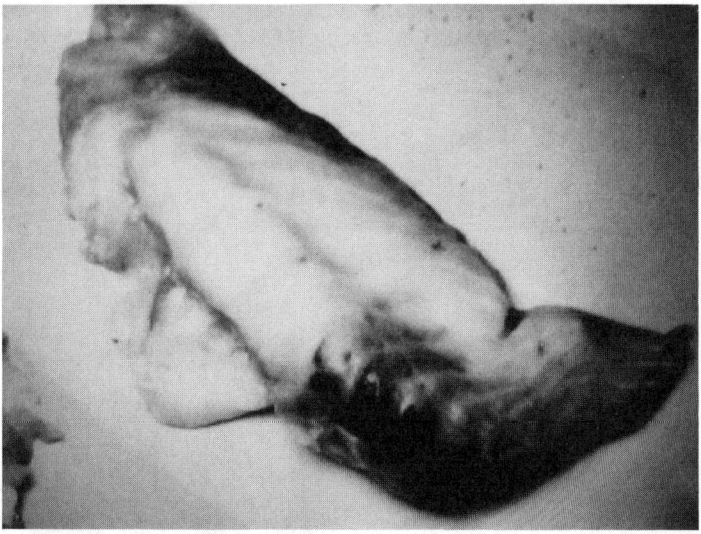

Figure 22 The picture of the fish fillet with a red filter in front of the camera, taken from the monitor system. Note that the blood spots are not detected.

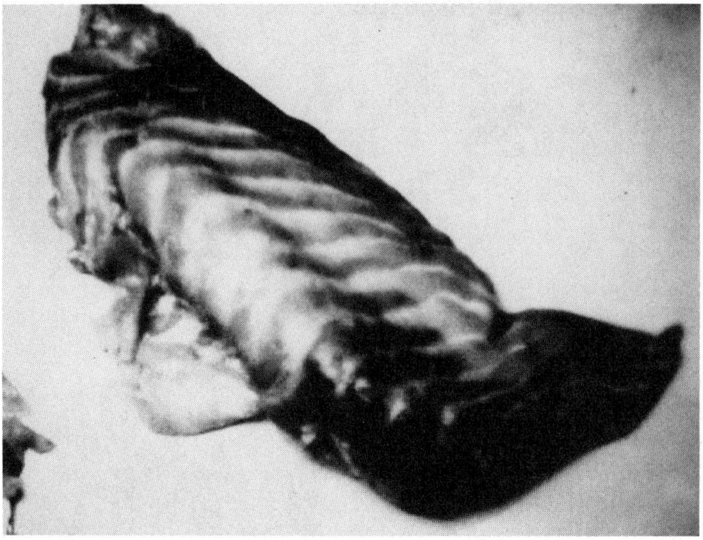

Figure 23 The picture of the fish fillet with a blue filter in front of the camera, taken from the monitor of the system. Note that the system detects flesh as a blood spot.

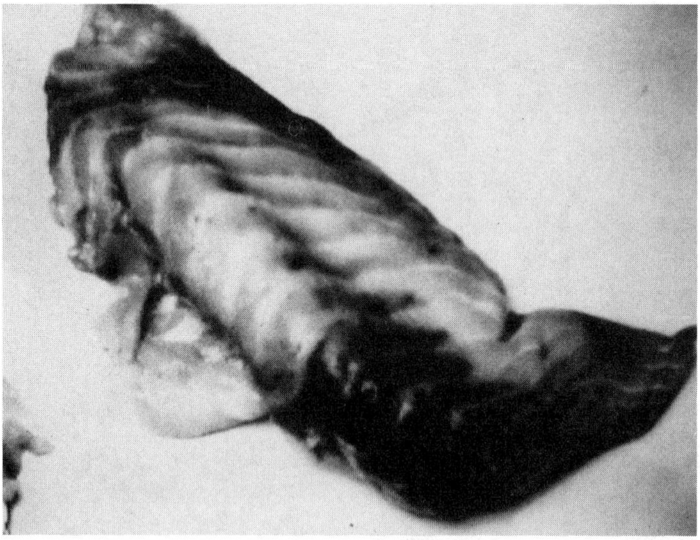

Figure 24 The picture of the fish fillet with a green filter in front of the camera, taken from the monitor of the system. Note that the detection of blood spots is good.

Table 3 Optimum Filter

Fish sample	Defect	Remarks	Lighting	Environment	Filter	Contrast
#1	Ventral skin	Grayish reflecting skin (lighter part)	Discharge lamp	Alternative method 1	Downgreen	Good
#2	Ventral skin	Grayish reflecting skin (darker part)	Discharge lamp	Alternative method 1	Downred	Bad => Green down Red top
#3	Blood spots	Small blood spots	Discharge lamp	Alternative method 1	Downred	Bad => Downgreen
#4	Blood spots	Bad brownish flesh	Discharge lamp	Alternative method 1	Downgreen	Good
#5	Discoloration	Bad brownish flesh	Discharge lamp	Alternative method 1	Downgreen	Good
#6	Discoloration	Lighter brownish flesh	Discharge lamp	Alternative method 1	Blue down Green top	Bad => Downgreen

- BlRe. Blue filter in front of the camera, red one in top of the lamp
- GrBl. Green filter in front of the camera, blue one in top of the lamp
- GrRe. Green filter in front of the camera, red one in top of the lamp
- ReBl. Red filter in front of the camera, blue one in top of the lamp
- ReGr. Red filter in front of the camera, green one in top of the lamp

The testing was done objectively and subjectively—Subjectively taking pictures of the monitor of the vision system. The results show that the best picture is obtained for the green filter in front of the camera (Figs. 21–24). That result is in agreement with the theory.

Objectively, we looked at the average pixel value of windows of 8×8 pixels size from different samples of fish. The results are shown in tables 4 and 5 (see Appendix II).

From the values of Tables 4 and 5, Table 3 is obtained. In that table it is shown that for method 1b—spatial light distribution, with the discharge lamp Osram HQL-R 125 de Luxe—the best results (such as better contrast between the transport belt and the flesh, better contrast between the flesh and the ventral skin and better contrast between the flesh and the blood spots) are achieved with the green filter in front of the camera. That result is in agreement with the theory and the subjective testing.

3. CONCLUSION

The recommended configuration will be:

- *Spatial distribution of light*: Front lighting according to Section 2.1, and method 1b of Section 2.1.
- *Temporal distribution of the light*: Stroboscopic light with a flash source according to Section 2.2. Bauer Ultrablitz 38 Logic according to the Appendix IV.
- *Spectrum of the radiation illuminating the object*: Discharge lamp according to the Section 2.4.
- *Spectral response of the lenses*: BK7 or SF11 lenses according to Section 2.5.

APPENDIX I

The following diagram shows the transmission spectrum characteristics of the red, blue and green filters.

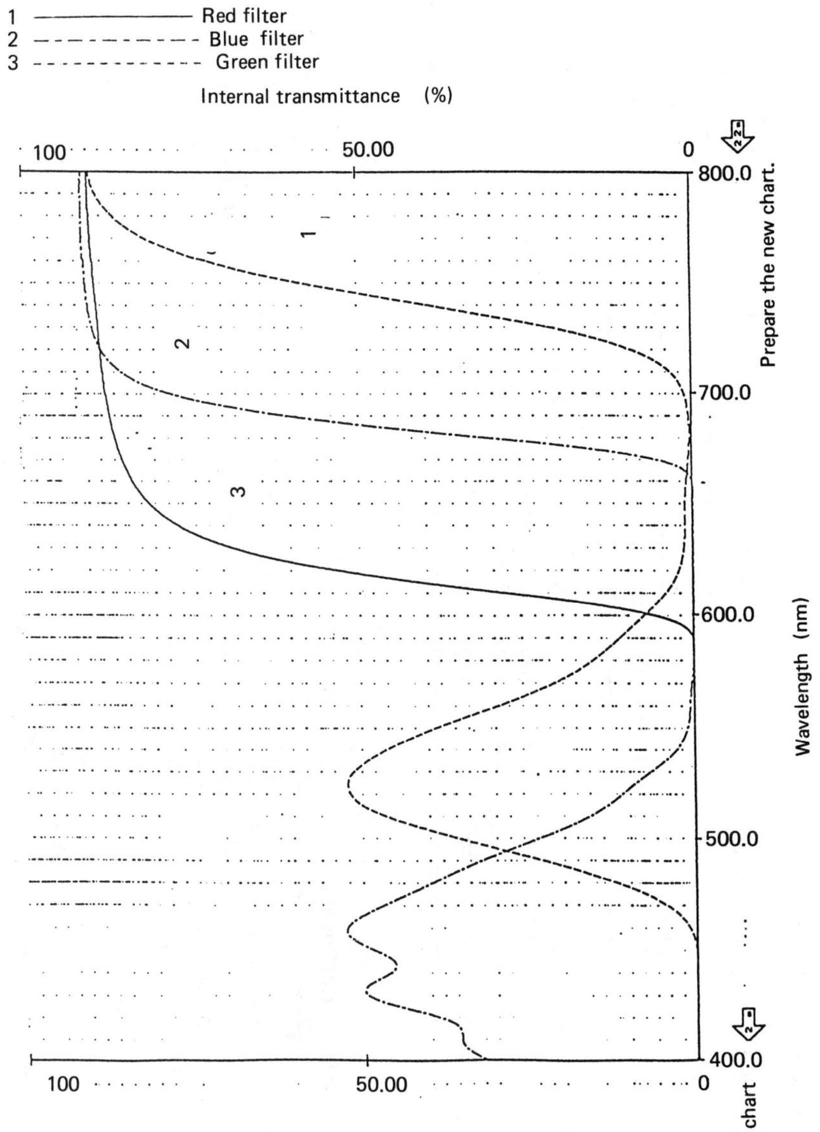

APPENDIX II

Table 4 Gray Level Readings in Windows 8×8

		Nothing	Down			Top			Combination					
			Blue	Green	Red	Blue	Green	Red	BlGr	BlRe	GrBl	GrRe	ReBl	ReGr
Fillet 1	#1	180	179	172	200	186	186	191	174	194	185	174	210	211
	#2	147	155	149	144	155	155	157	145	153	154	144	154	148
	#3	151	141	145	155	155	157	158	134	149	155	146	170	167
	Flesh	195	192	198	213	198	195	203	176	201	204	194	217	218
	Band	255	255	255	255	255	255	255	255	255	255	255	255	255

Table 5 Gray Level Readings in Windows 8×8

		Nothing	Down			Top			Combination					
			Blue	Green	Red	Blue	Green	Red	BlGr	BlRe	GrBl	GrRe	ReBl	ReGr
Fillet 2	#4	94	109	99	120	103	98	101	112	119	103	104	120	109
	#5	111	122	107	134	114	113	119	119	134	109	108	128	120
	#6	209	205	206	249	207	203	208	186	217	214	213	253	261
	Flesh	239	249	240	254	233	231	235	244	253	247	247	265	255
	Band	248	255	255	247	248	243	254	255	255	255	255	255	255

APPENDIX III

The adjacent program was developed for the ITEX 200 Interpreter of the SERIES 200 image processor from Imaging Technology Inc. The program obtains the average pixel value of one 8 $\times$ 8 window in one image stored before as 256 $\times$ 256. It is possible to change the size of the window going inside the program. *Note*: The prompt * indicates parts of the program that can be changed by the user.

```
i0 = 0;
x1 = (x coordinate value of the starting point of the window); *
y1 = (y coordinate value of the starting point of the window); *
dx1 = (horizontal size of the window);                         *
dy1 = (vertical size of the window);                           *
hdx = 512;
hdy = 512;
vdx = 256;
vdy = 256;
x = 0;
y = 0;
offset = 0;
depth = 8;
aoi0 = 10;
aoi1 = 20;
string = 0;
media = 0;

basicItexInitialize( );
aoiCreate(i0, x, y, hdx, hdy, depth, offset, aoi0);
aoiCreate(i0, x, y, vdx, vdy, depth, offset, aoi1);
createWordBuffer(string, 1);
aoiRestore(aoiRestore,"(name of the file to restore)");        *
display(aoi1);
```

Appendix III (Continued)

```
for (i=x1; i<x1+dx1; i+=1) {
    for (j=y1, j<y1+dy1; j + = 1) {
        aoiCreate(0, i, j, 1, 1, 8, 0, 20);
        aoiReadVerticalLine(20,0,string,FALSE);
        media=media+readWordBuffer(string,0);
        printInteger(readWordBuffer(string,0);
        print("\n");
    }
}
print("\n");
printInteger(media/(dx1*dy1));
sleep(10);
exit( );
```

The program when running, gives us the pixel value of each pixel of the 8 × 8 window and finally the average pixel value, keeping this value for a short time.

APPENDIX IV

Table 6 shows the technical specifications of two types of flashes: the standard photographic one and the Bauer Ultrablitz 38 Logicflash.

Table 6 Technical Specifications of Two Types of Flash

		Standard	Bauer Ultrablitz 38 Logic
Price (Dkr) (1989)		400–500	1800–2000
GN (at ISO 100)		20	38
Aperture range (F–stop)		5.6	2.8/4/5.5/8/11/16
Variable power output		No	1/1–1/4
Automatic distance range		—	0.4–56 m
Flash duration			
Maximum		1/1200	1/1000
Minimum		1/25000	1/30000
Recycling time			
Manual	Batteries	8 sg[a]	7 sg
	NC accumulator	No	3 sg
Computer	Batteries	No	0.2–7 sg
	NC accumulator	No	0.2–3 sg
Winder		No	0.2 sg

Considering that the speed is a very important parameter in image-processing applications, it is very important to note the paramount importance of the recycling time, so the Bauer Ultrablitz 38 Logic will be used to freeze the fish fillet motion.

[a]sg = recycling cycles per second.

REFERENCES

1. Batchelor, B. G., 1985. *Lighting and Viewing Techniques, Automated Visual Inspection*, IFS Publications, Bedford, England, pp. 102–179.

2. Novini Amir, 1986. *Fundamentals of Machine Vision Lighting*, SPIE, Vol. 728.

3. Novini Amir, 1987. Fundamentals of strobe lighting for machine vision, *Conference Proceedings Vision '87*, Society of Manufacturing Engineers, Detroit, Mich., pp. 4–13 - 4–25.

4. Harding, K. G., 1987. Optical considerations for machine vision *Lighting & Optical Tutorial, SME Vision '87*, June 8, 1987, Detroit, Mich.

5. Melles Griot, *Optics Guide 3*, 1985.

6. OSRAM, *Light for inside and outside*, 1987.

8

Illumination Techniques for 3-D Machine Vision

ALASTAIR McLEOD *Center for Industrial Research, Oslo, Norway*

1. Introduction
   - 1.1 3-D Machine Vision
   - 1.2 Classification of Techniques
2. Illumination Methods for 3-D Vision
   - 2.1 Moiré Topography
   - 2.2 Lidar
   - 2.3 Structured Light
3. Structured Light in Detail
   - 3.1 Basic Principles
   - 3.2 Coherent and Incoherent Light Sources
   - 3.3 Methods of Generating Structured Light
   - 3.4 "Intelligent" Structured Light
   - 3.5 Detection, Analysis, and Range Calculations
   - 3.6 Calibration, Precision, and Accuracy
   - 3.7 Speckle, Shadowing, and Other Problems

4. Applications of Structured Lighting in the Fish Industry
   4.1 Linescan Camera System
   4.2 Area Camera System
5. Conclusions
   5.1 The Potential of Structured Lighting Techniques
   5.2 Direction of Further Work
   Appendix
   References

1. INTRODUCTION

Machine vision is commonly used to automate measurement and inspection tasks. In the fish-processing industry, detection of fish orientation, detection of bones and parasites, head removal, cutting, filleting, and dimensional measurement are of particular interest. Due to the variety of tasks, many different machine vision techniques will have to be employed to meet these requirements.

The topics considered below are methods of illumination for three-dimensional (3-D) machine vision, which are suitable for position and measurement tasks.

1.1 3-D Machine Vision

There is an increasing demand for 3-D machine vision. New methods of obtaining 3-D information have been developed in the past few years to add to established techniques. Both old and new methods are being taken out of the laboratory and onto the factory floor, making 3-D imaging the area of fastest growth in machine vision (1).

The imaging techniques described below measure the distance between the sensor and points on the surface of an object. This type of imaging is called topographic and, strictly speaking, is not fully three-dimensional. Some refer to it as 2 1/2-D imaging. A true 3-D image acquisition technique would produce data on points in a volume. (These points are called volume elements or voxels.) X-ray tomography is an example of such a technique. For the purposes of machine vision, however, it is convenient to refer to topographical methods as 3-D imaging.

There are many ways of acquiring 3-D information from an object or scene, and even more ways of processing and analyzing the results. 3-D images can be obtained using multiple sensors (e.g., stereopsis), structured lighting, interference techniques (including holography and Moiré imaging), and range finding by time-of-flight methods. 3-D information may also be obtained from a two-dimensional image using: stadimetry techniques (shape-from-shading, range-from-size); prior knowledge of a scene; 3-D modeling and other methods (2).

1.2 Classification of Techniques

There are several ways of classifying 3-D techniques according to the method of acquisition or analysis. One such scheme is shown in Figure 1. The classification of a particular technique is less important than an understanding of the basic principle behind it, including its advantages over other methods and its limitations.

STADIMETRY	• Shape from shading • Range from size
MULTIFRAME	• Stereopsis • Kinesis, optical flow • Multiple illumination & frames
STRUCTURED LIGHT	• Triangulation • Space encoding • Light striping
INTERFERENCE METHODS	• Interferometry • Holography • Moiré • Speckle
OPTICAL RADAR	• Time of Flight (scanner) • CW phase detector • Chirped FM

Figure 1 A classification list of 3-D vision techniques.

It does help, however, when assessing the range of available methods and the interrelation between the physical principles.

Almost all of these techniques acquire the depth* information by use or manipulation of an illuminating source. Stereopsis is one exception to this, where two cameras are used to form one image from two slightly different angles of view. Stereopsis, structured light, and Moiré techniques all get their depth information from triangulation calculations in one form or another. In contrast, lidar systems obtain their depth information by direct measurement from the sensor to the object.

2. ILLUMINATION METHODS FOR 3-D VISION

Most of the techniques listed in Figure 1 obtain the depth information by use or manipulation of a particular illuminating source. In this section, however, only the techniques considered most suitable for real-time applications in the fish industry are considered.

2.1 Moiré Topography

Moiré techniques were used for surface contouring long before modern image processing techniques were developed (3). The advent of digital-imaging techniques has widened the range of application, and although few industrial measurement or inspection systems yet employ the methods, it has been successfully applied to many measurement problems in the laboratory.

Applications of Moiré topography include measurement of stress in steel panels and contouring of turbine blades.

There are many variations on the Moiré technique, but the basic principles are similar. Figure 2 illustrates the projection method, where a striped pattern or grating is projected onto the object. A second similar grating is placed in front of the sensor (usually a camera), which produced an interference or beat pattern. The observed sensor image is, therefore, a series of light and dark areas or fringes called Moiré fringes. These fringes represent contours of equal height on the object, although the direction of height variation is not implicitly defined.

*As is common practice, within the machine vision industry, the terms depth, height, and range are used interchangeably, and all refer to the distance between the image sensor and a point on the image scene.

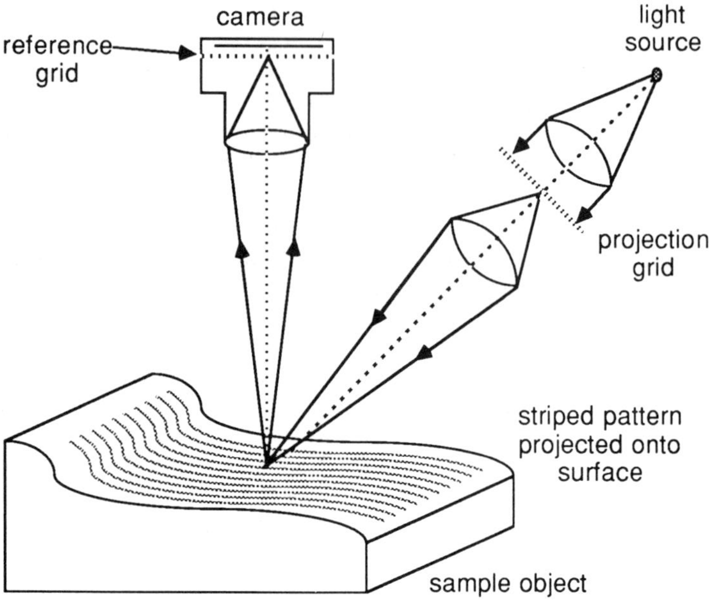

Figure 2 The Moiré projection method.

A slight variation on this method is a shadow technique (Fig. 3). In this case, only one striped grating is used, through which both the light source and the camera view are directed. The grating combines with its shadow to produce the fringes in the camera image. Alignment problems are minimized with this method, but the grating must be placed close to the object, and there are limitations on the object size. Moiré topography has the potential for very high precision measurements, and height resolutions of around 0.1 mm can be achieved with both of these methods.

One further way of using a single grating is to project the pattern onto the object and calculate the reference pattern in software, after the image has been digitized and stored. This method is called computed Moiré and results in a simpler optical system at the expense of increased processing complexity. Its resolution is lower than the projection or shadow method, because the camera must resolve the grating pattern itself rather

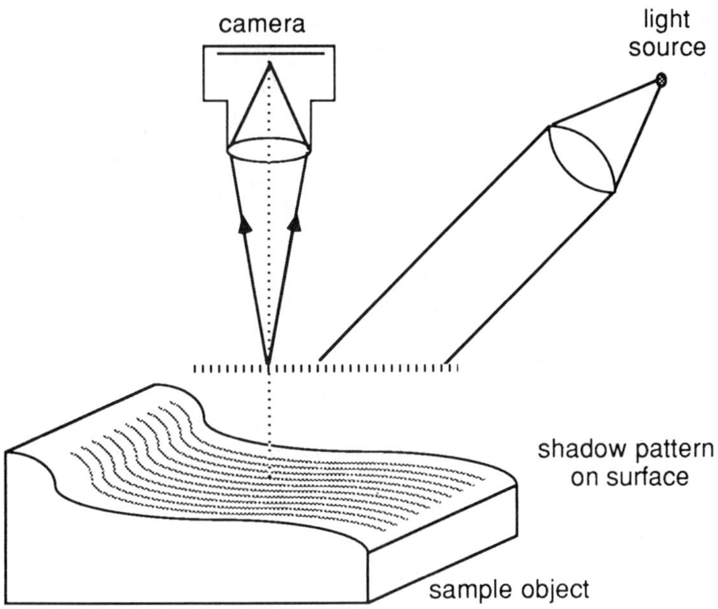

Figure 3 The shadow Moiré method.

than just the fringes resulting from the interference of two grating patterns.

A major problem in the application of Moiré to industrial inspection is the difficulty of interpreting the fringes. The fringe image must be scanned and analyzed to extract the depth information. Sophisticated software and a powerful computer system are required to complete these tasks at an acceptable speed for industrial machine vision purposes. In many cases direct structured light methods may be more effective and less cumbersome to implement (see Sec. 3).

One very interesting recent development, however, is the interpretation of Moiré interference patterns using an optical image processing technique (4). Here the Moiré fringes are analyzed in Fourier "feature space" using an Optical Fourier Feature Extractor (OFFE), developed by Global Holonetics Corporation. With this device, the Moiré pattern of curves and lines is converted into two Fourier feature signatures—one representing line orientation and one representing line spacing or frequency. This signature is obtained instantly, as it is an optical process, and allows rapid comparison with signatures previously obtained. In this way, the 3-D differences between similar objects can be detected. If a range of signatures representing "good" or acceptable objects are stored, then these can be compared with each new signature obtained and an accept or reject decision can be made.

This technique is still under development, but promising results have been obtained in the inspection of such objects as tin cans, plastic bottles, and sheet metal.

To summarize, Moiré techniques offer high accuracy in depth measurement, but present systems have some limitations and are probably not suited, or too sophisticated to apply, to the particular problems of the fish-processing industry.

2.2 Lidar

Lidar, an acronym for light detection and ranging, was initially developed long before the advent of the laser, and is sometimes referred to as optical radar. With the laser, however, lidar techniques rapidly improved, and were applied to military range-finding, surveying, and air-pollution monitoring (5). Electronic imaging is one of the latest areas of application of lidar, made feasible by the development of compact laser systems, high sensitivity sensors, and fast electronics.

The basic principle of lidar is simple, and is identical to that of ultrasonic methods and conventional radar. The time interval between the transmission of a light pulse and its return after reflection off a target is directly proportional to the target range due to the constant speed of light.

Measuring range points directly by this pulse echo method avoids some of the problems associated with other methods of 3-D image acquisition, such as obscuration and positional ambiguity. In the past, pulse-echo has been rather more suited to longer range measurements, i.e., ten or more meters up to several kilometers. This method has not been successfully applied to much shorter range measurements.

Over distances of up to a few meters, a range resolution of around 10 mm can be obtained (6). Using averaging techniques, which of course will dramatically reduce range acquisition speeds, resolutions of 0.1 mm are feasible. This requires careful design of the electronic timing system (7).

A more sophisticated approach employs a cw (continuous wave) laser beam modulated at high frequency as the light source (8). With this method, the difference in phase between the transmitted and received signals is measured, rather than the time of flight of a single pulse.

The modulation frequency is the determining factor of range resolution. The higher the frequency, the shorter the modulation wavelength, and therefore shorter distances can be resolved. Only the phase angle of the returning wave is known, and not the absolute number of wavelengths. There is an uncertainty, therefore, termed the ambiguity distance, of one half of the modulation wavelength. For instance, if the modulating frequency were 20 MHz, the wavelength would be 15 meters. The uncertainty in the absolute range would be 7.5 meters. That is, two points 7.5 meters apart would produce the identical phase angle signal. Over this distance it would be possible to resolve 30 mm or less. Again, averaging techniques can improve the system performance at the expense of speed.

The laser ranging methods described above measure range at one particular point on the object. Obviously, to obtain a full range map of a surface, many points must be measured. This is usually achieved by reflecting the laser off one or two scanning mirrors, moving the measurement point across the object. If the object is on a moving conveyor belt, just one mirror component is used to scan the laser in one dimension. Two are used to scan the laser beam in both X and Y directions. (Laser scanning is discussed further in Sec. 3.3).

2.3 Structured Light

The term "structured light" refers to illumination of objects by a projected pattern of light with a predefined geometry, to extract depth or range information. Structured light is the most widely used form of topographical imaging in machine vision. In general, structured lighting is used in machine vision applications to detect object presence and location and to measure height at specific points. The projection of a single well-defined plane of light is sometimes called "light sectioning," as it generates part of the outline of a cross section of the object intersecting it.

Theoretically, the technique is well suited to the problem of fish measurement. It is also likely to be less expensive to implement and easier to develop into a working system then Moirè or lidar methods.

For this reason the discussion in the next section is devoted entirely to structured lighting.

3. STRUCTURED LIGHT IN DETAIL

This section reviews the basic principles of structured lighting and describes methods of generation, methods of image detection and analysis, limitations, and examples of industrial applications.

3.1 Basic Principles

Structured lighting is a noncontact gauging technique employing the triangulation principle. The simplest form of structured light is the optically unaltered beam from a cw laser, which projects a point of light into the surface of an object. Figure 1, each position along the laser beam is imaged onto a unique position on the sensor. This position is given by:

and

$$M = \frac{\Delta d}{\Delta D}$$

and so

$$\Delta d = \Delta h . \tan \Phi$$

Thus the height of the object at the laser spot is directly proportional to the position of the spot image on the sensor. This simple principle is readily extended into two dimensions by replacing the spot with a plane of light. This is the most commonly used light "structure" or pattern. It is conveniently achieved with

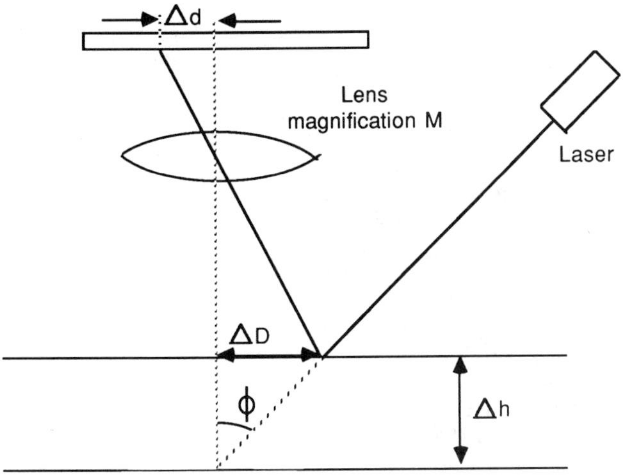

Figure 4 Laser spot triangulation.

$$\tan \Phi = \frac{\Delta D}{\Delta h}$$

a laser and a cylindrical lens (Fig. 5). An area sensor then detects the position of the line at each point over its length, giving a height profile or contour of the object (Fig. 6). A contour map of the surface is generated by sweeping the laser stripe over the object or by moving the object through the light plane.

3.2 Coherent and Incoherent Light Sources

Both coherent and incoherent light sources can be used to generate structured light patterns. Coherent light sources are most often used to generate simple light patterns such as spots, lines, or cross hairs. Projection of complex patterns with coherent sources can give rise to unwanted interference effects, and incoherent sources are usually preferred in these cases.

Halogen lamps, xenon arc lamps, and fluorescent strips are the most common sources of incoherent light. Arc lamps, due to their cost and relative complexity, are only employed when very

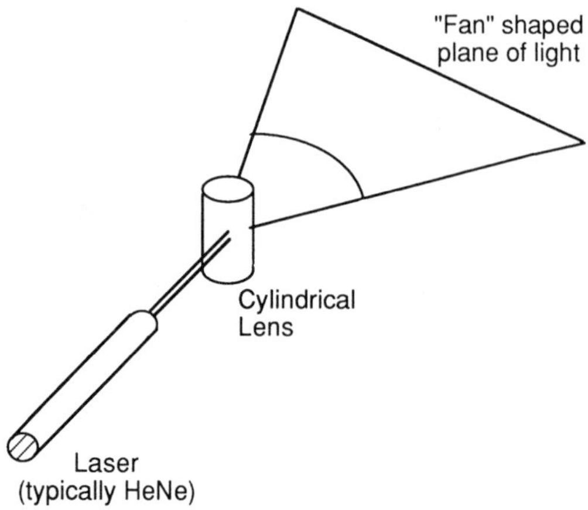

Figure 5 Laser stripe generation using a cylindrical lens.

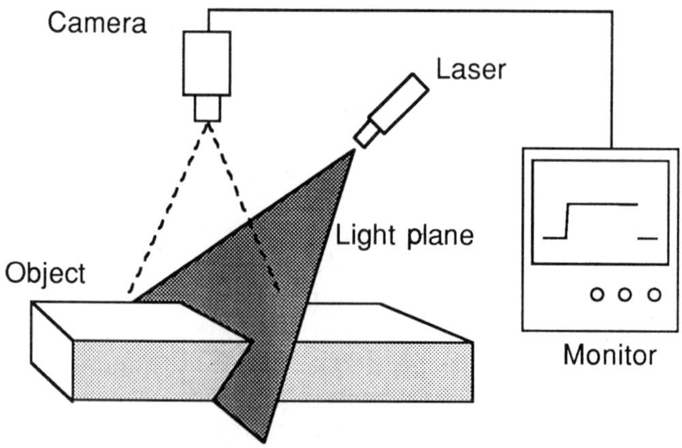

Figure 6 Laser "light sectioning".

high brightness is required. Halogen lamps, in common with arc lamps, generate quite a lot of heat, which may be a problem in some situations. Higher power halogen lamps (50–150 W) are usually needed to achieve sufficient contrast in a projected light pattern. Fluorescent strips are commonly used for machine vision illumination but not for structured light generation due to their large source size. As the output from this type of lamp varies rapidly, in phase with the power supply voltage, they must be driven with a higher than normal AC frequency. This allows the camera to average out the lighting variations within each frame period.

Coherent sources, i.e., lasers, are easier to "shape" into the desired pattern than incoherent sources. The quality of the focused light pattern is usually very much better with coherent sources. It is not too difficult to reach a diffraction-limited performance on a focused beam (i.e., the focus is limited only by diffraction effects and not by the optical aberrations of the beam-shaping optics).

A cw laser output of 5–20 mW is usually sufficient to gain acceptable results.

There may be occasions when the use of a particular wavelength of light source is required due to some special properties of the object. Cost considerations in applications of structured lighting will often limit the use of different laser types to just helium-neon lasers (632.8 nm) and near infrared laser diodes (750–900 nm).

Green (543 nm), yellow (594 nm), and orange (612 nm) helium-neon lasers are now available at reasonable cost, and laser diodes are becoming commercially available in shorter and shorter wavelengths down to about 670 nm (see Appendix). This allows some choice in the decision of which wavelength of laser to use.

3.3 Methods of Generating Structured Light

For convenience, the described methods of producing structured lighting are divided into coherent and incoherent systems.

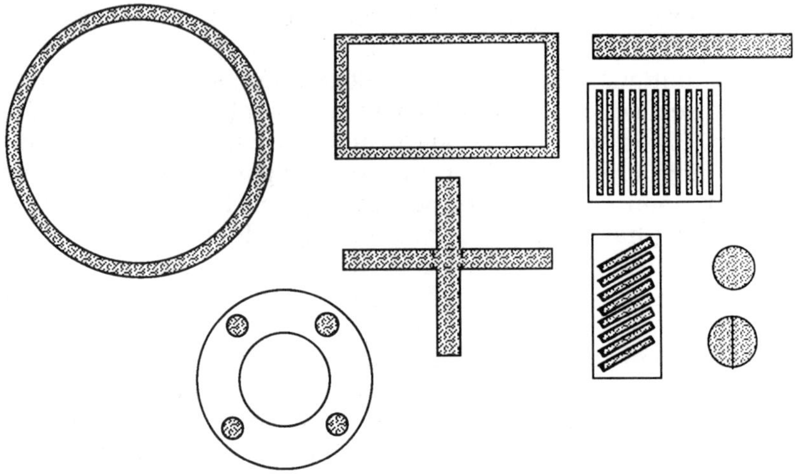

Figure 7 Fiber-optic bundle end shapes.

Incoherent (White Light) Sources

1. *Spot/stripe projection*: A simple spot or line may be projected using white light sources. A pinhole or slit is placed between the source and a collecting lens. The beam is collimated and projected on to the object.
2. *Slide projection*: Using the slide projector principle, the desired pattern is produced on a standard 35 mm slide and then projected and focused onto the object. In some cases, it may be sufficient to use an ordinary slide projector adapted for the purpose. The distribution of projected light, however, will not be uniform and can cause problems, especially if simple thresholding techniques are used to process the camera image. It is generally better to use a special collimated white-light projector, designed for the purpose (see Appendix).
3. *Fiber-optic cross section*: Using what is known as a cross-section converter, a fiber-optic bundle can be adapted to form simple end-shapes (Fig. 7), such as lines, circles, crosses, etc. With a suitable lens, these can be imaged onto the object, creating the desired light pattern. Consideration must be given to the resolution of the

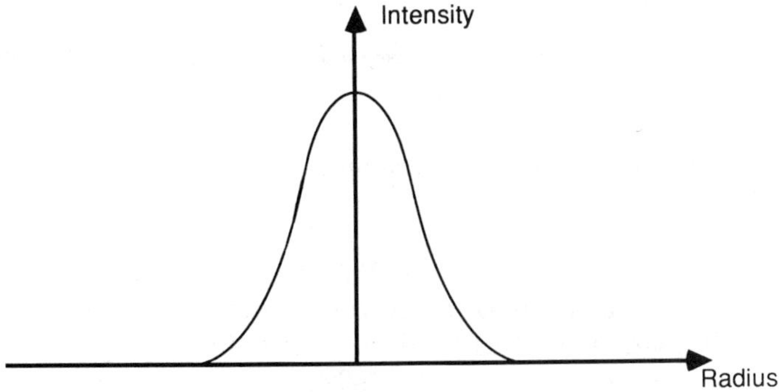

Figure 8 Gaussian laser beam cross section.

camera and lens system, and the magnification (or demagnefication) of the fiber cross-section image so that the ends of hte individual fibers are not resolved in the final digitized image.

Coherent (Laser Light) Sources

1. *Spot projection*: An unmodified laser beam provides an ideal spot projection source. If a spot size is required which is different from the raw laser beam output, the beam must be expanded using special optics and, if required, refocused onto the target. Working range and depth of focus must be carefully considered.
2. *Cross hairs*: Projector optics are available which produce a small cross hair pattern from the laser beam. This can be useful in certain circumstances.
3. *Stripe*: The most common pattern of light "structure" or pattern is a single straight line. This is often produced using a section of glass rod placed in front of the laser beam. This method is not good enough if quantitative measurements are to be made on the stripe position. The intensity variation produced perpendicular to the projected line should be Gaussian, as depicted in Figure 8. Careful design of the optics is required to reduce distortions in the projected line, such as asymmetry and "bow-tie" broadening (Fig. 9), where the line be-

Figure 9 "Bow-tie" broadening of laser stripe.

comes wider at its ends. This is essential if subpixel calculations are to be attempted. Fortunately, lenses specially designed to produce well-defined projected lines are available for both visible and near infrared laser sources (see Appendix). These lenses have an adjustable focus to allow for variation in standoff distance and either a fixed or adjustable divergence angle. Lenses with adjustable divergence are equivalent to camera zoom lenses and are often termed "zoom line projectors."

4. *Scanned beam:* A laser line or point can be scanned across a surface. Laser beam scanning can be accomplished by spinning mirrors (e.g., rotating polygon mirrors), resonant scanners, galvanometer scanners, acousto-optic cells, and spinning holographic disks (Holotek). Sweeping a laser stripe across a surface is one method of obtaining a full range map of stationary surface. If a laser spot is scanned in a line or in an area pattern (such as a raster scan), synchronization between the scanning system timing and the frame rate of the imaging sensor must be considered carefully. Accurate surface mapping using 3-D scanning is often quite slow, taking several minutes for a complete scan, although much faster systems have been built (8).

3.4 "Intelligent" Structured Light

A variation on laser-scanning systems for machine vision is the concept of dynamic or intelligent structured light. A computer controlled *x-y* laser scanner system produces a scanning pattern on an object, which is sensed by the imaging system and analyzed by the computer. The computer can then adjust the direction of scanning, the shape and size of the scan pattern, and even the focus of the laser spot to suit the object being scanned. This method of adaptive scanning can concentrate on areas of interest on the object by increasing the scan pattern resolution, for example.

Overall system speed may be increased significantly by first carrying out a fast, rough scan, supplemented by accurate scans of detailed areas, and ignoring large flat areas.

Work on intelligent structured light is currently being carried out by the author and his colleagues at SI.

3.5 Detection, Analysis, and Range Calculations

A single, well-defined projected laser stripe is likely to be the most suitable light structure for measurements of fish samples. The discussion in this section is therefore mainly related to the detection and analysis of a single laser stripe setup as shown in Figure 6.

The stripe pattern is imaged onto the camera sensor by the camera lens. A narrow optical bandpass filter, which transmits the laser wavelength only, can be used to reduce noise and sensitivity to changes in ambient light levels.

The camera signal is captured by a video frame grabber system and displayed on a monitor (Fig. 6). Assuming that the geometry of the camera and laser are known, the height of each point on the laser stripe at each point can be simply calculated by applying Eq. (1), as it is just an extension of the single spot case into two dimensions.

For processing simplicity, the laser line is usually projected parallel to the horizontal or vertical axis of the camera. The intensity of the stripe cross section should also be Gaussian (see Sec. 3.3). To further reduce noise, the camera signal can be thresholded at a suitable level. The selection of threshold level is very important, and should be made at the point where the slope of the Gaussian intensity profile is a maximum. Ideally, the output level of the illuminating source and/or the sensor exposure level should be adjusted to suit object reflectivity to produce the best results.

The range calculations remain simple if the camera and laser angle are fixed and the object is moved through the structured light pattern (on a conveyor belt, for example). If the laser

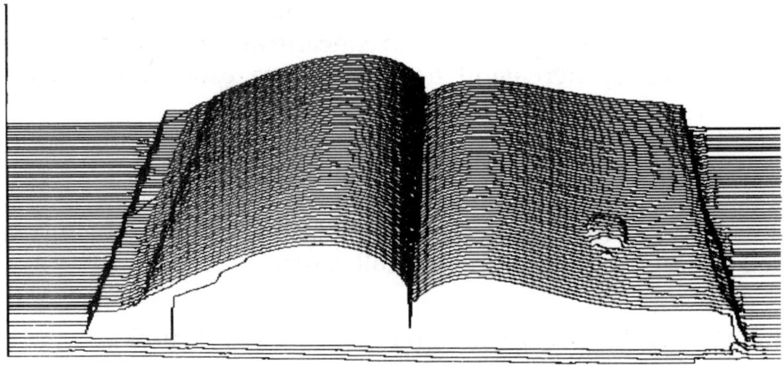

Figure 10 3-D profile map of book and coin.

angle changes to sweep a line or point over the surface, the range calculations become slightly more complex as both the height and the gap or step size between successive lines becomes a function of the laser angle.

The surface data can be displayed as a series of profiles to give a 3-D representation on paper or monitor screen. Figure 10 shows such a profile display. This image of a book with some coins placed on one page was obtained by an X-Y laser scanning method (see Sec. 4.3 for further details). Another method of display is to map each height value to a grey level and show the variation in surface height of the object as a grey level image.

3.6 Calibration, Precision, and Accuracy

For many applications (including fish measurement), a good quality camera lens will provide sufficiently high performance so that calibration to correct optical distortion is not required. For accurate dimensional measurement, calibration is a necessity (9). Optical distortions can produce positional errors of up to several pixels towards the edges of an image (10). Distortions in the laser stripe can be dramatically reduced by using a high quality line projector lens, as discussed in Section 3.3.

Calibration of the system removes sources of systematic errors and increases the measurement accuracy. Precision is always limited by random errors and in this case is basically a

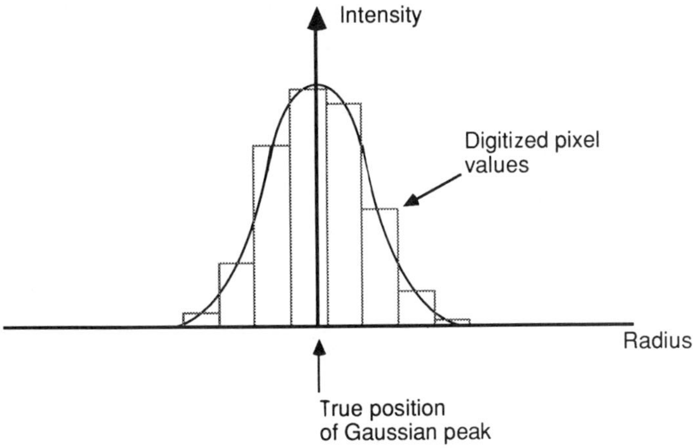

Figure 11 Subpixel calculations of Gaussian peak position.

function of the system resolution and signal noise. Height resolution in the setup shown in Figure 6 is not only a function of camera resolution and laser profile definition, but also of laser angle. Increasing the angle between the sensor and the laser plane increases height resolution, as a fixed change in height produces a larger change in laser position. If the laser stripe cross section is Gaussian, the center of the stripe (i.e., the Gaussian peak) can be calculated to a greater accuracy than that given by the raw pixel resolution (Fig. 11). Subpixel calculation techniques allow increase of precision, but must be used carefully, with a thorough understanding of the process involved.

The pitfalls of subpixel calculation techniques are well covered elsewhere and do not concern us here. The application of machine vision to the fishing industry is unlikely to require high precision.

3.7 Speckle, Shadowing, and Other Problems

Speckle is a random distribution of intensity which is seen by the eye or other sensing system when coherent light is reflected from an unsmooth surface. Speckle destroys the Gaussian profile of the laser beam, rendering subpixel calculations inaccurate. In

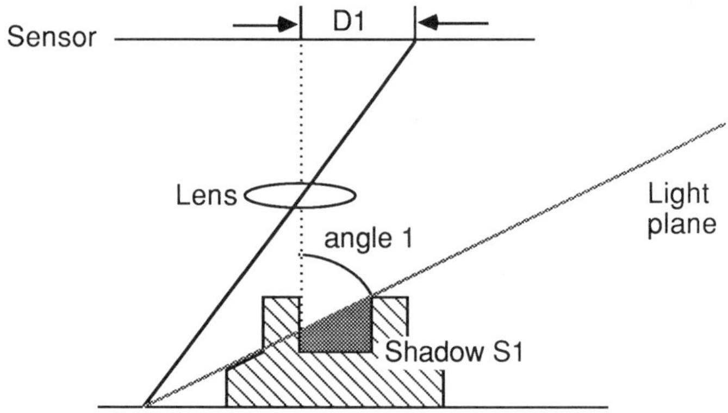

Figure 12 The "occlusion" or shadowing problem.

applications where high precision is not required, this can be neglected.

All triangulation-based range sensors have limitations due to occlusion or shadowing. There are areas which are shadowed from the projected light pattern by the surface geometry of the object and are therefore not mapped by the imaging system. This condition is illustrated in Figure 12.

As mentioned previously, height resolution increases with projection angle, but the area in shadow also increases. There is, therefore, a trade-off between resolution and occlusion, which is illustrated in Figures 12 and 13. In addition, depending upon the reflection properties of the surface, increasing the laser angle is likely to reduce the amount of light scattered towards camera.

Variation in surface reflectivity can cause serious difficulties in acquiring range data for a complete surface. Increased camera dynamic range and sensitivity can help to overcome this problem. On some surfaces, such as machined or extruded metal, reflections of the laser stripe can produce spurious lines in the image (Fig. 14). Prior knowledge of the object type can allow the line-analysis algorithm to ignore such reflections.

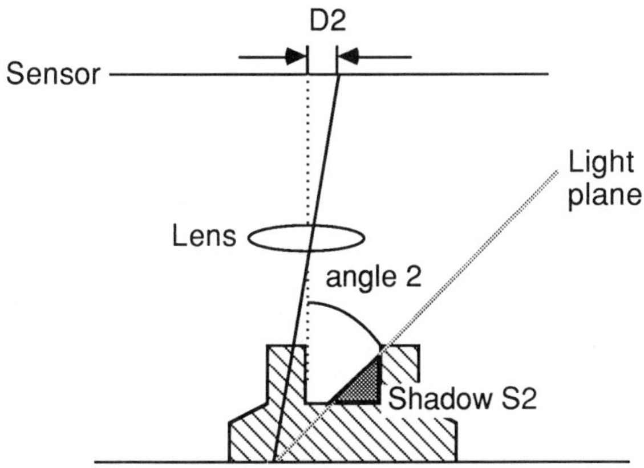

Figure 13 The trade-off between occlusion and resolution.

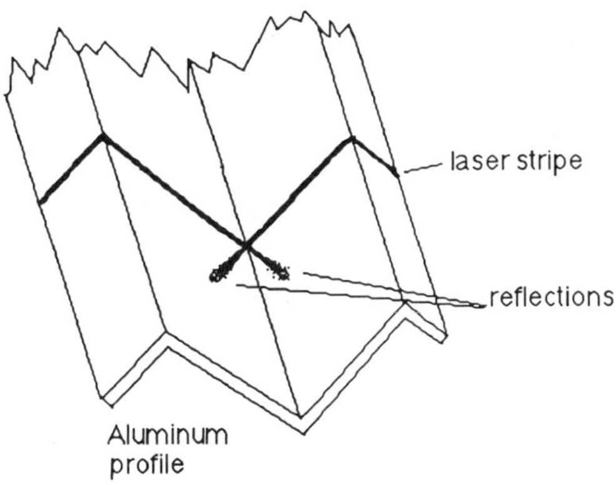

Figure 14 Reflection of a laser stripe in an extruded profile.

Some surfaces also exhibit a definite grain structure. Examples are turned, ground, milled, and extruded surfaces. The orientation of the grain direction can affect the accuracy of structured light sensors. It is wise, therefore, to determine experimentally the relative orientation of the camera, surface, and projected light pattern that produces the best results.

4. APPLICATIONS OF STRUCTURED LIGHTING IN THE FISH INDUSTRY

In this section the concept of structured light is considered when applied to measurement of fish shape and size. Preliminary experimental results for both systems are discussed.

4.1 Linescan Camera System

New types of linescan cameras are available with preprocessing facilities built into the camera itself. These so-called "smart" cameras have standard serial or parallel outputs and can be directly connected to a PC or other host computer system, without the need for a specialized video interface. The cameras are preprogrammable via this interface. Data such as edge point position, edge type (positive or negative), and distance between edges is output by the camera. Simple thresholds or gradient levels can be pre-set. Scan speeds can be varied to suit lighting levels and required data acquisition rates.

Such sensors could easily be applied to measurements of outline and orientation of fish on a conveyor belt. In ideal circumstances, this technique would be simple to implement and would not require the use of structured lighting. Real-world circumstances may not be ideal, however. If the reflection of the background surface (e.g., the conveyor belt) cannot be reliably controlled, contrast is likely to be a problem. Given the natural black and silver fish coloring, contrast may well be a problem anyway (see below). One solution may be to use a projected laser stripe to aid the linescan sensor.

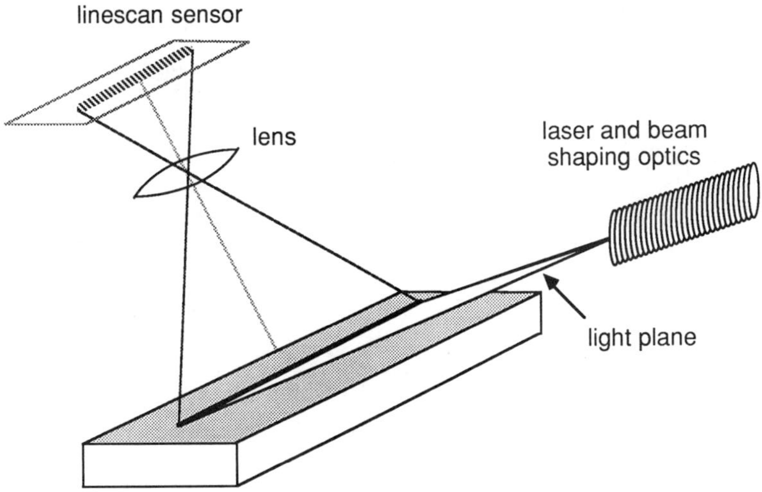

Figure 15 Linescan array viewing laser stripe.

Experimental Work

Consider the experimental setup shown in Figure 15. A laser stripe is projected at an angle onto the belt. The linescan camera is positioned directly above the moving belt with the linescan sensor viewing the length of the laser stripe. When no object (e.g., a fish) is present, the camera will view an unbroken line. As a fish passes through this line (Fig. 16), the line becomes broken as the light strip intersects the fish, out of the line of sight of the linescan sensor. The gap between the negative and positive edges of the resulting camera signal is a direct measure of the width of the fish at that point. If the conveyor belt speed is constant, an accurate outline of the fish can be built up by the host computer. From this, the position and orientation of the fish can be calculated and used to direct an automatic cutting machine, for example.

Figure 17 shows a reconstruction of the experimental data obtained by moving a herring through the light beam. (As no proper link was made to a host computer, the measurement points were noted manually from the liquid crystal readout display supplied with the linescan camera (Pulnix model IMS-2k).

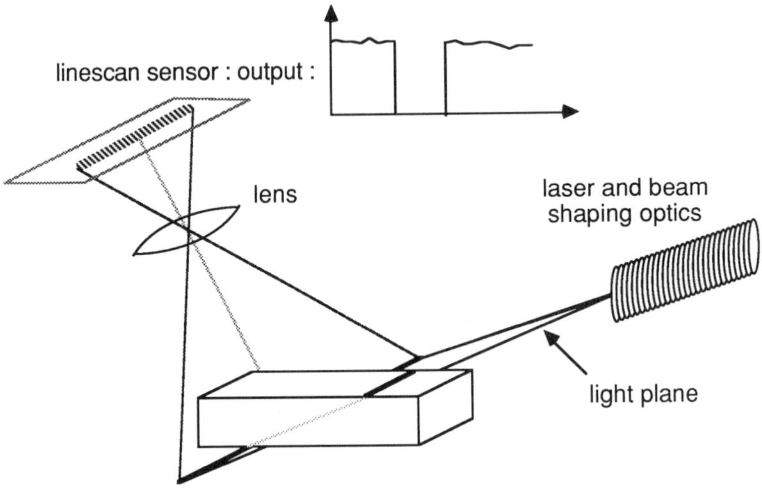

Figure 16 Shape of laser stripe intersection and linescan output.

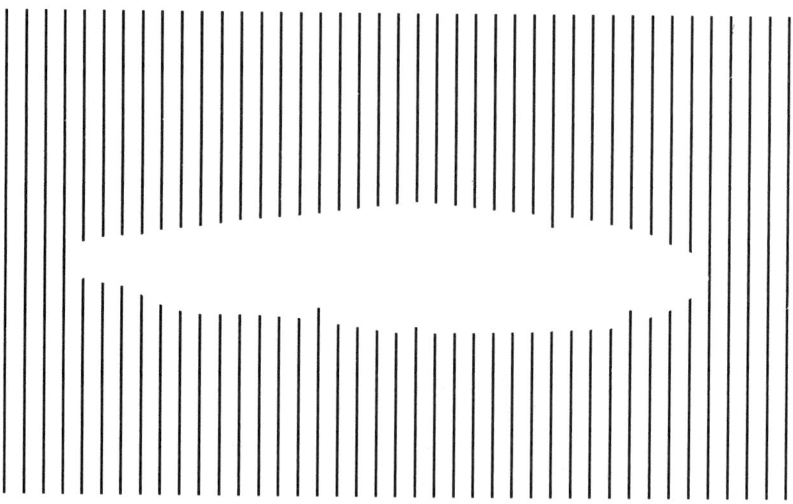

Figure 17 Reconstructed linescan outline of a whole herring.

The camera was supplied with a Pentax 50mm f1.8 lens with a focus friction-lock. The laser used was an NEC 5 m W HeNe model, fitter with a Newport Corp. zoom line projector lens.

The system is inexpensive, fast, and accurate. Smart linescan sensors are available from several manufacturers (see Appendix). The sensor need not have a dedicated host computer, but could be interfaced with a central system which controls other tasks in parallel.

4.2 Area Camera System

By replacing the linescan camera with an area camera in the system described above, the height of the fish at each point along the laser stripe can be measured. The size, shape, and volume of the fish can be calculated from repeated measurements as the fish passes through the beam (assuming its speed of movement is known).

This method could be used to measure the size of a whole fish or to position a cutter to remove head and tail. Data from measurements of filleted fish could be used for the accurate division of fillets into equal volumes and for determining an optimum cutting strategy.

Experimental Work

In this experiment, the laser stripe itself was created using a computer-controlled X-Y scanner, allowing a plane of light to be scanned across the fish sample, thereby simulating a moving conveyor belt. A new image was captured for every line position, and a profile map constructed, clearly illustrating the way in which such a technique could be used to measure fish volume.

The principle of measurement has already been described in some detail above, and so the matters of interest here are the experimental factors which might determine the limits on performance of such a system. No volume calculations have yet been made, as these are merely preliminary experiments to demonstrate the feasibility of the principle and determine possible difficulties or drawbacks.

General Comments

Three types of fish were scanned: whole herring, gutted sea trout, and cod fillet. As one might expect, the dark upper sides of both the trout and herring reflected the laser beam quite poorly, while the silvery undersides reflected it very efficiently. What was surprising, however (at least to the author), was the degree to which the dark upper side absorbed the light. A matte black test surface reflected the light many times more strongly than the fish itself.

In addition, the tail fins were quite transparent, and some of the areas around the jaws were translucent, giving a glowing effect over a large area. This was especially true of the cod fillets, all of which glowed brightly under the laser scanner. By carefully adjusting camera lens aperture and grey level thresholds, encouraging results were obtained despite these difficulties.

The end result of these effects was that parts of the fish were not always imaged accurately. It would be of interest to use both longer and shorter laser wavelengths to see if better results could be obtained.

Results

Figure 18 shows a pair of whole herring in a profile scan representation. The heads are pointing towards the camera viewing posi-

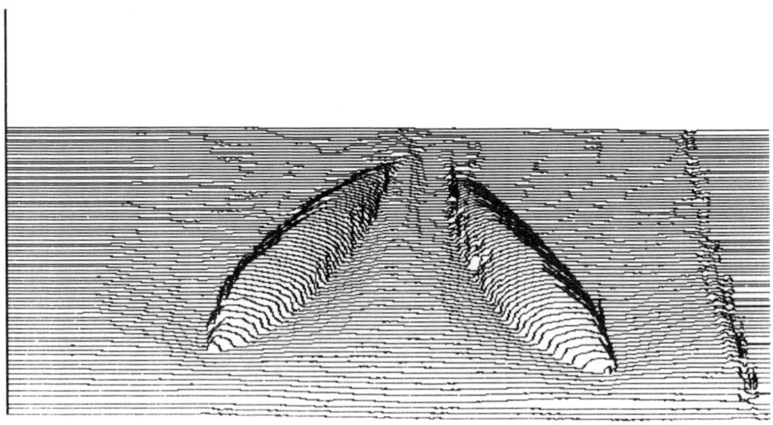

Figure 18 3-D profile map: herring.

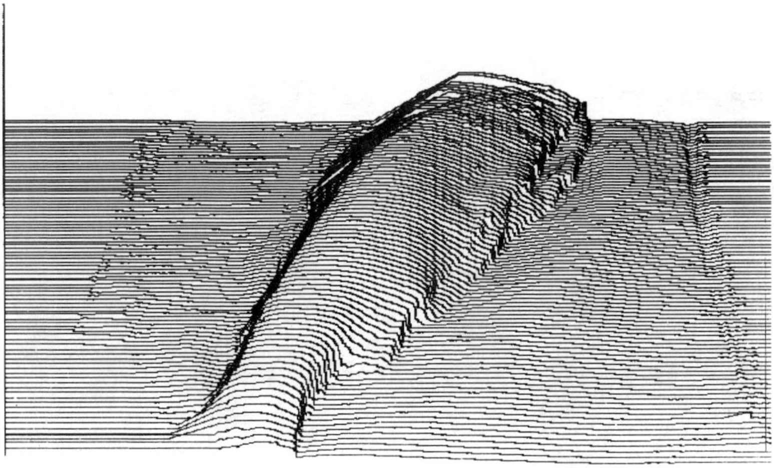

Figure 19 3-D profile map: sea trout body.

tion. A higher magnification camera lens would have provided more detail in this instance. The dark sides of the fish are both towards the outer edges of the scan. The fish were placed on a paper surface, which began to flex slightly around the fish, due to their moisture content. This effect can just be discerned in the profile map, particularly in the area between the two fish.

A large (~2 kg) sea trout was tested using the same experimental layout. The resulting profile map is shown in Figure 19. The sea trout was whole, but had been gutted. This time the head is pointing away from the camera, and in fact lies outside the scanning field. The cut in the underside of the body which had been made when gutting the fish can clearly be seen on the right side. Due to the size of the fish, some parts on the left side are in shadow and are not mapped properly.

The final profile map, shown in Figure 20, was made with a higher magnification lens on the camera. The fish sample was a piece of filleted cod, placed with the skin side on the table. Although the laser produced a glow throughout the flesh of the fish and the light was scattered unpredictably, a good result was obtained. Although fish volumes were not calculated from these results, preliminary estimations indicate that fish volume could be measured with an accuracy of 2–3% or better. (*Note*: The occa-

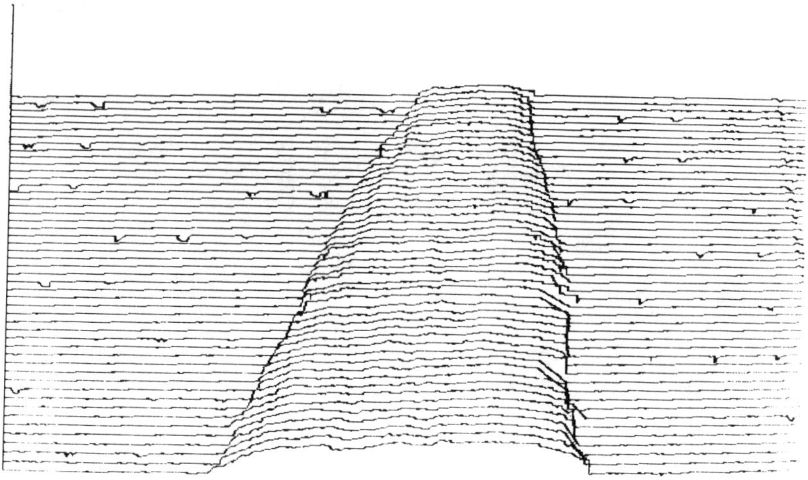

Figure 20 3-D profile map: cod fillet (higher magnification).

sional blips in the profile map in Figure 20 are due to holes drilled in the optical table on which the fish was placed.)

5. CONCLUSIONS

Selecting the most suitable 3-D system requires matching the demands of the application with the performance characteristics of the machine vision system.

For systems based on structured light, such parameters as depth resolution and accuracy, reflectance sensitivity, working range, occlusion angle, and measurement speed must be considered. Cost, complexity, development time, environmental considerations, ease of use, and system size are of course additional factors common to any machine vision application.

5.1 The Potential of Structured Lighting Techniques

Structured light is a simple technique and an effective method of acquiring 3-D surface data. It is a suitable method for measurement of fish size and volume. It does not need to be too expensive to implement in terms of hardware costs and could be implemented with off-the-shelf components.

The smart linescan sensor system described would require little development and would be inexpensive. It has limited application, however, as only fish outline and position would be measured.

The area camera system offers a simple way of measuring volume, employing relatively unsophisticated image-processing techniques. The software development time is therefore likely to be shorter than with other more complex methods. The range calculations are based on simple trigonometry, and so processing speeds could be high. Hardware processing to extract the line data is feasible and may be cost effective if many systems were required.

Occlusion is unlikely to be a problem with fillets and smaller fish, but the situation may be different with larger fish.

Structured light has high potential may be one of the most useful machine vision techniques to apply to the fish industry.

5.2 Direction of Further Work

The experimental results described were carried out to test basic principles and were not optimized in any way. They show that structured light can be used for measurements of fish and that further investigation would be useful. A clearer picture of the requirements of the fish industry is necessary to develop such systems in the optimum way.

It is clear that the results of the present experiments could be improved by several factors, including the use of a more sensitive camera, a higher power laser source, and more advanced image analysis software. It may also be possible that the wavelength of the laser source may be better matched to the optical properties of fish. A logical next step is to develop a volume calculation algorithm and test the measurement system against calibrated targets.

ACKNOWLEDGMENTS

The author would like to thank his colleagues Øivind Dahle and Bjørn Lilliekjendlie for their work in programming the image processing routines and the laser scanner drivers, and also for their help with some of the illustrations. The Pulnix linescan camera was kindly loaned to us for testing purposes by Holund A/S, Oslo. The fish used for the experiments were not wasted, but were much appreciated by both the author's family and their neighbor's cat. (Perhaps scanning by laser is an essential part of a novel cooking recipe?)

APPENDIX A: LIST OF EQUIPMENT SUPPLIERS AND PRODUCTS

Newport Corporation, Fountain Valley, California
Line projector lenses, zoom line projector lenses, HeNe lasers, visible and near infrared structured light modules, optical mounting components, camera lenses, calibration targets, optics, colored filters, laser bandpass filters, collimated white light projectors.

Melles Griot, Rochester, New York.
Optics, optical mounting components, colored filters, laser bandpass filters, green HeNe lasers, near infrared and visible laser diodes, laser diode line projector modules.

Ealing Electro-Optics, Watford, UK.
Optics, colored filters, laser bandpass filters, collimated white light projectors.

Oriel Corporation, Stratford, Connecticut
Optics, colored filters, laser bandpass filters.

PMS Electro-Optics, Boulder, Colorado.
Green, yellow, orange, and red HeNe lasers.

Toshiba, Tokyo, Japan.
Visible laser diodes.

NEC, Tokyo, Japan.
Visible laser diodes, HeNe lasers.

Pulnix, Sunnyvale, California.
CCD area cameras, linescan cameras, smart linescan cameras.

Allen Bradley, Milwaukee, Wisconsin.
Smart linescan cameras, vision systems.

Honeywell Control Systems, Bracknell, UK.
Smart linescan cameras, range sensors.

General Scanning, Watertown, Massachusetts.
Laser scanning systems, resonant scanners, galvanometer scanners.

REFERENCES

1. West, P. C., 1987. 3-D Imaging complexity of machine vision, Advanced Imaging, August.
2. Shirai, Y., 1987. *Three-Dimensional Computer Vision*, Springer-Verlag (New York).
3. Reid, G. T., 1984. Moiré fringes in metrology, Opt. and Laser Eng., 5: 63–93.
4. Belilove, J., and R Hekker, 1988. A new Moiré Fourier transform inspection system, *SME Vision '88 Conf. Proc.*
5. Beesley, M. J., 1978. *Lasers and Their Applications*, Taylor & Francis Ltd.
6. Ahola, R., R. Myllylä and A. Kilpelä, 1986. A fast time-of-flight distance sensor for industrial applications *SPIE Proc.*, Vol. 654, Aut. Opt. Insp., p. 12.
7. Ahola, R., and R. Myllylä, 1986. A new Method for measuring the time-of-flight in fast laser range finding, *SPIE Proc.*, Vol. 654, Aut. Opt. Insp., p. 19.
8. Hershman, M., F. Goodwin, S. Kenyon, and A. Slotwinski, 1987. Coherent laser radar application to 3-D vision and metrology, *SME Vision '87 Conf. Proc.*
9. Dahle, Ø., A. McLeod, and A. Sommefrelt, 1989. Automatisk system for inspection av ekstruderte profiler, Norges Teknisk-Naturvitenskaplige Forskningsråd (NTNF) report No. SI–87 01 20–1.

10. Butler, D. A., and P. K. Pierson, 1987. Correcting distortion in digital images, *SME Vision '87 Conf. Proc.*

IV

Remote Sensing

9

Fishery Applications of Satellite Remote Sensing and Image Processing in the Gulf of Mexico

Rex C. Herron *National Marine Fisheries Service, Bay St. Louis, Mississippi*

♦♦♦

1. Introduction
2. Tuna Research
3. Menhaden Research
4. Mississippi River Plume Research
5. Gulf Butterfish Research
6. An Overview and Outline of Current Technology for Obtaining AVHRR Satellite HRPT Images on Vessels at Sea

References

♦♦♦

1. INTRODUCTION

The National Marine Fisheries Service (NMFS) utilizes image processing in its remote sensing studies of living marine resources. The majority of our remote sensing research in the Gulf of Mexico (GOM) utilizes sensors aboard spacecraft in polar orbits. The satellites utilized most often include the Nimbus–7 polar orbiter containing the Coastal Zone Color Scanner (CZCS), from which chlorophyll concentrations and the location of sediment-laden waters can be estimated, and the NOAA TIROS-n polar orbiters containing the Advanced Very High Resolution Radiometer (AVHRR), from which sea surface temperatures (SST) are estimated. AVHRR High Resolution Picture Transmission (HRPT) data provide the most useful images with the greatest flexibility and quality. The AVHRR also detects in a broad visible band that is useful for some applications.

AVHRR sensors aboard the NOAA TIROS-n meteorological satellites yield SST estimates to within about 0.5°C (1). Several NOAA polar orbiters are now operational, and coverage of an area can be obtained several times during a 24-hour period. Occasionally, thermal images of SST from the NOAA geostationary weather satellite are used in fisheries applications, although ground resolution restricts their uses.

The CZCS became operational in 1978 and stopped functioning in mid-1986. It detected in blue, bluegreen, green and red visible bands, one near infrared band and one emitted infrared band, and provided temporal coverage approximately 2 of 5 days (2). Chlorophyll pigment concentrations, indicated by an increase in green and a decrease in blue radiances, are generally estimated to within a factor of two to three. Another ocean color sensor is tentatively scheduled for launch on a satellite platform in the early 1990s by the United States (3).

Occasionally, data from the LANDSAT multispectral scanner (MSS) are used in fisheries investigations. These images yield a spatial resolution of less than 100 m in two visible and two near infrared bands (2). Use of these images is severely restricted in fisheries research because of their infrequent (once about every 18 days) and small areal (185 × 185 km) coverage, however, ground resolution enhances their attractiveness for some studies.

The temporal and spatial distributions of chlorophyll, other suspended solids, and temperature can provide much information about the system investigated. Fronts, areas of increased vertical and horizontal mixing, are formed where water masses of sufficiently different physical and biological properties abut and are important habitat features for many marine organisms. Fronts can be detected with satellite thermal or color imagery, and the intensity of the horizontal temperature gradient can be calculated from thermal imagery. This satellite-derived parameter has proven useful in studies in the GOM (4). Other oceanic features, such as streamers and eddies, can be discerned. These types of water masses are currently under investigation to determine their impact on marine ecosystems. For example, eddies regularly form in the DeSoto Canyon area of the northeastern Gulf, and their impact on larval fish survival and subsequent recruitment is under investigation (5). Water mass motion can be deduced from a time series of satellite images, and numerical models to predict direction and speed of water masses from AVHRR thermal images are under investigation. Some studies are described in greater detail.

2. TUNA RESEARCH

NMFS observers, placed aboard Japanese long line bluefin (*Thunnus thynnus*) and yellowfin (*t. albacares*) tuna fishing vessels in the 1978 (March through mid-July) fishing season in the GOM, compared vessel locations with selected physical parameters and satellite-measured ocean thermal and color characteristics (6). Nearly all Japanese captains apparently used SST in some manner as a guide to fishing strategy, although it was not clear whether they were seeking a particular isotherm, horizontal temperature gradient, or a combination of both.

Historical mean monthly temperatures suggested strong horizontal temperature gradients at or near the estimated hook depths during tuna long line fishing. Satellite GOES (8 km ground resolution) thermal imagery (weekly composites) showed the major continuous surface temperature front to be the edge of the warm Gulf Loop Current. The fishing fleet tended to locate their sets to the outside or near the edge of the Loop Current, and the distribution of catches was related to the Loop Current (7). By early June, heating of the ocean surface resulted in a near isothermal condition in the Gulf preventing satellite detection of important features such as the Loop Current. During this period, the Japanese switched from fishing for bluefin to yellowfin tuna in the western Gulf.

The GOM is a known spawning area for bluefin tuna in the western Atlantic Ocean. Spawning generally occurs from mid-April to mid-June throughout the northern and eastern Gulf. Feeding and spawning requirements probably influence the distribution of adult and larval bluefin. Upwelling of nutrient-enhanced deep waters often occurs in areas of geographic relief, along the edge of the Gulf Loop Current, and along the edges of Loop Current extensions into cooler Gulf waters. Upwelling also occurs in streamers of shelf water and in cold core eddies entrained by the Loop Current.

Many of these features can be detected with satellite imagery. Studies, initiated in 1987, used satellite AVHRR thermal imagery to locate larval fish sampling transects on frontal boundaries at the edge of the Gulf Loop Current and at edges of other mesoscale features. Larval bluefin tuna were caught only at or near the Loop Current boundary and in cold core eddies, and none was caught within the main body of the Loop Current (Fig. 1). Of the captured bluefin tuna larvae, 90% were caught within 20 km and on the cool side of the Loop Current boundary near where the 25°C isotherm intersected the surface.

Total fish larvae concentrations were highest near the Loop Current boundary and on the cool side (Fig. 2), as were total plankton concentrations as evidenced from plankton displacement volumes (8,9). Many larvae were unidentified yolk-sac larvae, however, larval Myctophidae (lanternfishes) dominated in

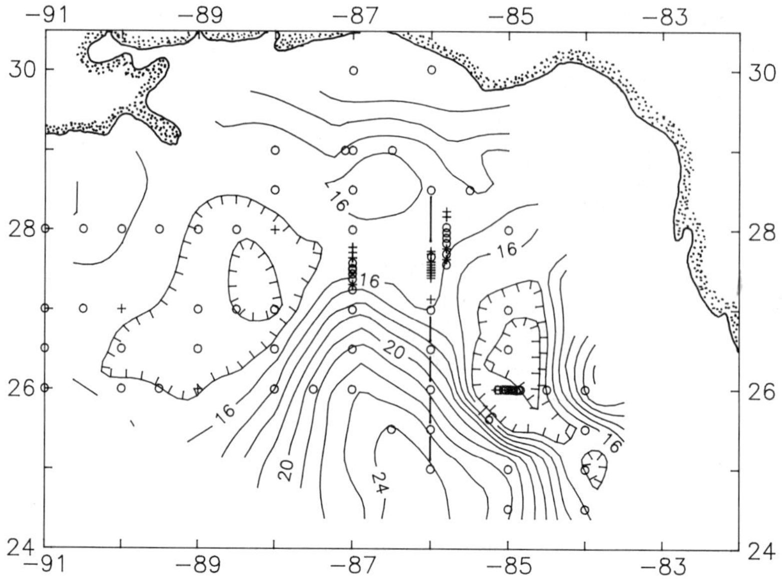

Figure 1 Temperature (°C) at 200 m depth for May 7–22, 1987 in the Gulf of Mexico. Crosses indicate stations where at least one bluefin tuna larvae was caught, and circles represent all other sampled stations. Solid line indicates vertical section locations along 86° W. Longitude (9).

frequency and abundance. Diversity and species richness were higher near the Loop Current boundary.

One might be tempted to conclude that the only favorable habitat in the GOM for larval bluefin tuna is the cool edge of the Loop Current or in some sold core eddies. Other hypotheses, however, deserve consideration. These areas may be preferred by adults for spawning for an as yet unknown reason. Physical mechanisms, such as convergence, may act to concentrate larval bluefin tuna, or predation in these areas may be reduced. These hypotheses need further testing, and satellite imagery will play an important part in these activities.

We may be able to more precisely define the relationships between bluefin tuna and pelagic habitat from environmental and satellite data such that satellite images can be used to estimate the total annual available spawning and nursery habitat.

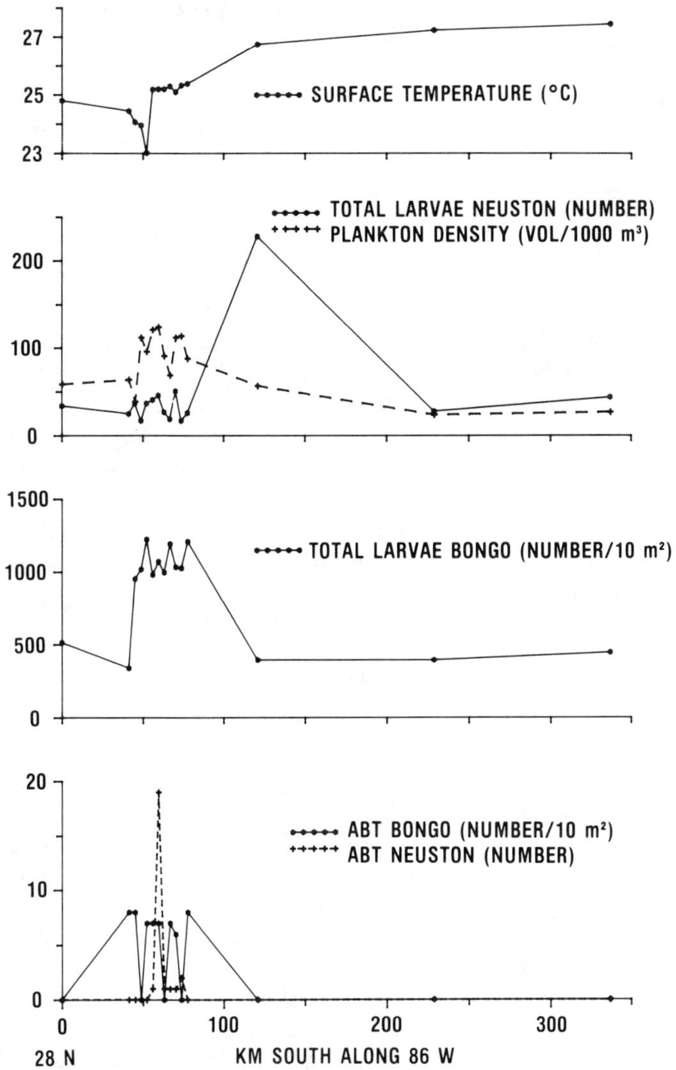

Figure 2 Comparison of sea surface temperature in the Gulf of Mexico and catches of larval fish and plankton along 86° W. Longitude for 400 km southward from 28° N. Latitude (9).

This may be incorporated into current methods for calculating indices of abundance from which bluefin tuna stocks are monitored. Improved indices of abundance would allow us to better manage the tuna stocks in the GOM by preventing overfishing and, hopefully, increasing the population size.

3. MENHADEN RESEARCH

NMFS participated in a project to analyze Gulf menhaden (*Brevoortia patronus*) fishing and satellite remotely sensed data collected from 1972 to 1975 along the northcentral coast of the GOM (10). LANDSAT satellites 1 and 2 provided multispectral scanner (MSS) radiance measurements in four bands. Locations and approximate sizes of menhaden schools were reported by spotters in fixed-wing aircraft, fishing vessel operators, and investigators aboard research vessels. Oceanographic data were collected on days of satellite coverage to develop algorithms for converting LANDSAT MSS data to estimates of turbidity and chlorophyll concentrations.

Radiance values in each spectral band, collected within 2 hours of fish observations, were compared in 10–15 areas with and without observed menhaden schools. Water turbidity and color were correlated with school distributions, and this information was used to construct a statistical algorithm to classify high probability menhaden areas from LANDSAT imagery. The results indicated classification accuracies between 80 and 90%. A probability chart, prepared from a LANDSAT image within 24 hours of the satellite flyover, was compared with menhaden locations reported by the fishing fleet. Most locations fell within or adjacent to the predicted high probability areas (Fig. 3).

The investigations in the mid-1970s did not lead to an operational system for predicting high probability menhaden locations because LANDSAT satellites, with repeat coverage of about once every 18 days, did not provide timely temporal coverage. With the launch of the CZCS, temporal coverage and reduced procurement and processing expenditures suggested a promising alternative. An analysis, similar to the LANDSAT analysis, was later performed (12). Raw radiance values from individual channels of the CZCS were used in a discriminant function analysis (DFA) to compare water color in areas where Gulf men-

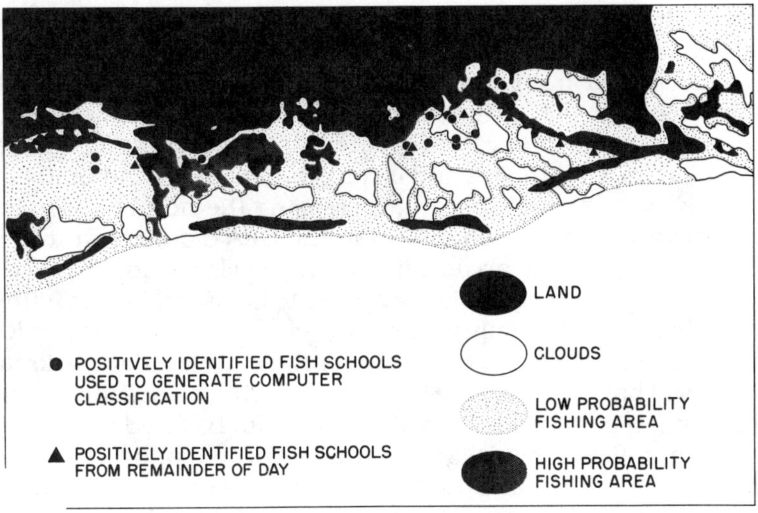

Figure 3 Gulf menhaden high probability areas predicted from a LANDSAT MSS satellite image and actual locations of observed schools in the northcentral Gulf of Mexico (11).

haden were caught (fish) with areas where menhaden were not caught (no-fish). The coastal region from Mobile Bay, Alabama, to western Louisiana was divided into grids of equal area approximately 12' of latitude and 10' of longitude to a side. Mean raw radiance values were calculated for each grid from pixel values taken from the CZCS during a spring and a fall orbit in 1981. Menhaden catch data were collected from spotters aboard aircraft and from commercial fishing vessels from within a few days of each orbit.

Mean raw radiances were obtained for 60 grids during the two CZCS orbits. Menhaden were caught in 51 of the 60 areas (fish) leaving nine no-fish grids. Mean raw radiance values from six channels of the CZCS were used in the DFA to compare the 51 fish with the nine no-fish areas.

Data from only two channels, channel 2 (the 520 nm bluegreen band) and channel 4 (the 670 nm red band), resulted in 96.7% correctly classified areas. Two of the 51 fish areas were incorrectly classified as no-fish areas, but all no-fish areas were correctly classified. Unstandardized discriminant scores (DS) were calculated from the equation: DS = −0.5672386(CH2) + 0.1241486(CH4) − 1.161846. DS values less than 1.25 were considered fish areas, and values greater than 1.25 were considered no-fish areas. Channels 2 and 4 contributed 49.96 and 50.04%, respectively, to the total score. Increasing the number of independent variables (i.e., channels) in the DFA did not result in a greater percent of the areas being correctly classified.

These types of analysis may prove to be useful in predicting areas where commercially viable schools of menhaden may be located. With the subsequent failure of the CZCS, an operational system will have to await the launch of another color sensor, or a fixed-wing, high altitude aircraft will have to be fitted with color sensors and flown regularly. A computer program, developed to calculate the DS for each pixel in a remotely sensed image, could produce charts representing areas of potential menhaden abundance. This could reduce search time spent by purse seining operations because more time would be spent searching for fish in predicted areas of high abundance. Conceptually, daily charts could be produced ready for immediate use.

4. MISSISSIPPI RIVER PLUME RESEARCH

AVHRR satellite imagery was used by NMFS to determine the boundaries of the Mississippi River plume for purposes of locating ichthyloplankton sampling stations. Satellite AVHRR visible images were used to detect the more turbid river plume when river water temperatures were similar to Gulf water temperatures and thermal imagery could not be used to distinguish the water masses. Early morning and late afternoon visible images were often obscured by sunglint, and a method of removing some sunglint (13), using information in the reflected infrared channel to remove much of the glint in the visible channel, was occasionally used to process the visible imagery.

Research vessels, directed to the plume boundaries using satellite imagery, sampled stations representing low turbidity Gulf waters outside the plume, plume boundary waters, and highly turbid waters inside the plume. Sequences of satellite images, collected approximately every 6 hours while vessels were on station, were used to constantly update the researchers on the location and disposition of the plume.

Primary production rates were higher, and surface chlorophyll concentrations were up to 20 times higher, in plume front waters than in plume or shelf waters (14–16). Copepod nauplii, an important prey item for larval fish, were more numerous near the river delta than in other regions of the GOM, and concentrations were usually correlated with chlorophyll concentrations (17).

Analyses of several years of historical data indicated that larval fish were more abundant in frontal areas associated with the plume than in areas away from the fronts (18). Data from the satellite-aided research indicated that larval and postlarval fishes tended to concentrate in the summer plume fronts, and larval Gulf menhaden caught in plume fronts usually contained more food in their guts and had faster growth rates than menhaden caught elsewhere. Over 30% of all larval and postlarval king (*Scomberomorus cavalla*) and Spanish mackerel (*S. maculatus*) caught since 1983 have been caught in the plume area. Young king mackerel, which feed almost entirely on larval fish, collected at the plume in 1986 grew significantly faster than those caught at other locations. Summer river discharge rates positively correlated ($r = 0.71$) with king mackerel recruited into the fisheries at age 2.

Several hypotheses have been proposed to explain these results. Physical convergence at fronts may tend to concentrate large planktonic organisms, and nutrients supplied by river water may stimulate high primary production at fronts with concomitant increase in zooplankton (19). Richardson (20) reported primary production rates in a North Sea frontal zone up to 25 times higher than in surrounding waters. Predators, such as mackerel, may concentrate at fronts to take advantage of enhanced feeding opportunities resulting from increased prey abundance (21). Remotely sensed data will be an important tool to aid researchers in testing these and other hypotheses.

5. GULF BUTTERFISH RESEARCH

Latent resources represent a large and potentially valuable commercial resource in the GOM. Gulf butterfish (*Peprilus burti*), coastal herrings, and associated species and squid constitute the greatest proportion of the latent resources potential in the Gulf. NMFS initiated a Latent Resources Research Program in 1983 to determine relationships between living marine resources and environmental factors, including relationships derived from satellite remote-sensing techniques. Gulf butterfish bottom stern trawling data, collected from survey and commercial cruises in the northern GOM, were used, in conjunction with relationships derived from satellite imagery (22,23), to develop and refine an expert system to be used in predicting locations of probable high butterfish concentrations (24). A catch rate of 454 kg/hr or higher was considered a large catch and was selected to represent a large school of butterfish.

AVHRR satellite SST images were collected in the springs of 1985–1987 within 3 days of trawling. A 3×3 pixel Sobel operator (25) was passed over the data in each SST image to produce weighted horizontal components of the temperature gradient. The computed output image was the gradient magnitude of the two horizontal vector components. Trawl locations were used to extract the horizontal temperature gradient magnitude at each site.

Catch rates were significantly correlated with SST gradients as were mean catch rates, determined by averaging catch rates for each resovable SST gradient value (26). The linear equation, Average Catch = −156.63+63.12*Gradient, was significant (n = 26, $r = 0.73$, $p = 0.0001$) when only one catch was made for a particular SST gradient value and that data point was deleted from the analysis. An exponential model eliminated the prediction of negative average catches at SST gradient values of 0. When SST gradients used in the analysis were taken only from images collected on the same days that trawls were made and no data points were eliminated from the analysis, an exponential model, Average Catch = 53.50*EXP(0.15* Gradient), resulted in an improved relationship ($r = 0.921$, $n = 23$, $p < 0.001$) (Fig. 4). Average SST gradients were significantly higher where butterfish were caught than at other locations.

Results from the butterfish SST gradient analysis were used, along with other environmental information, to construct a microcomputer-based prototype expert system (BUFERSS) for predicting locations of probable butterfish aggregations from

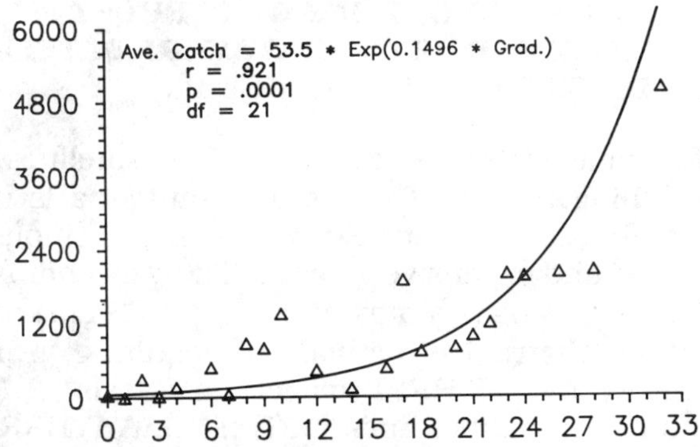

Figure 4 Average butterfish catch rates in the northeastern Gulf of Mexico as an exponential function of satellite image–derived sea surface temperature gradients using only catches made on the same day that satellite imagery was collected (26).

timely AVHRR satellite SST images (24). Butterfish distributions were related to bottom depth and temperature, SST, ocean color, and, possibly, moon phase in addition to SST gradients (23). Conceptually, the expert system analyzes a satellite image, along with environmental data, and produces a chart of probable butterfish locations and associated confidence factors.

Commercial fishing for Gulf butterfish was initiated in 1986, and the 1988 fishing season marked the third year of commercial activity targeting Gulf butterfish. Most commercial fishing had been conducted in depths of 20–175 fm during daylight hours in areas of previous success and where commercial quantities of butterfish had been caught during NMFS surveys. Trawling attempts made outside previously fished areas were initially influenced by the availability of satellite imagery information (4). When a thermal front(s) was detected with satellite imagery, vessel captains would be notified of the location of the front(s). Captains would proceed to seek out and fish the fronts. Vessel captains would report periodic successes from this technology and would frequently request updated satellite imagery information.

6. AN OVERVIEW AND OUTLINE OF CURRENT TECHNOLOGY FOR OBTAINING AVHRR SATELLITE HRPT IMAGES ON VESSELS AT SEA

AVHRR HRPT imagery is received directly from satellites on some larger, nonfishing vessels. The receiving antenna is unidirectional, and stabilization and tracking must be accomplished using complicated and expensive gyrostabilizing mechanisms that currently restrict its use on most vessels. NMFS currently uses and researches alternative methods of acquiring, processing, and transmitting AVHRR HRPT imagery to vessels.

Computer-compatible tapes (CCT) of NOAA AVHRR satellite HRPT imagery are currently available to NMFS, Mississippi Laboratories, from the U.S. Navy's receiving station at the Stennis Space Center (SCC) in southwestern Mississippi. There are no restrictions on the size of the image; the only limitation being the swath width of the scanner aboard the spacecraft. Benefits to NMFS of purchasing CCT's from a local source are several. The area of interest can be scrutinized for cloud cover within 30–60 minutes of the satellite flyover and prior to making an actual purchase. After making the decision to purchase, a CCT of the image is usually available within another 30–60 minutes. Requests for imagery of areas outside U.S. waters may be honored if arrangements for coverage are made in advance.

AVHRR satellite HRPT imagery can be acquired directly from the satellite with an earth receiving station. Land-based receiving stations are now available with unidirectional tracking antenna, receiver, and PC units capable of some image preprocessing. Costs of a complete system range up to about $200,000 (all currency in 1988 U.S. dollars) and may be available for as little as $100,000. One has virtually an unlimited source of satellite data, and control of the imagery remains fully in the hands of the receiving station operator. Receiving stations can be located anywhere, thus allowing service to any interested community.

Another option for obtaining NOAA AVHRR imagery involves placing orders for CCTs through NOAA/NESDIS located in Maryland. Knowing which areas are cloud-free and suitable for use puts a limiting constraint on the user, and the delay time between placing orders and delivery of CCTs can range up to several months. This method is not practical for providing immediate service.

An additional option for obtaining satellite imagery was made available to NMFS by NOAA's National Weather Service at the National Meteorological Center in Maryland in 1988. The National Weather Service has developed a service, which is available to qualified NOAA offices, to provide timely georegistered satellite SST images. At this time, the service is considered a research tool still in a development mode and is not a standard operational product. Georeferenced AVHRR satellite data are converted to 0.5°C SST on a mainframe computer and maintained for approximately 24 hours before being deleted from the system. The interested party connects to the system through a common carrier (telephone line) with a remote terminal and modem. Software allows the user to specify an area and orbit of interest within the past 24 hours. The size (up to a 512 × 512 pixel image), resolution, and central point is specified by the user, and a single-byte character file is processed and downloaded to the user's remote terminal.

Satellite imagery can be obtained rapidly with this method using a microcomputer (PC), modem, and telephone line. This makes satellite imagery potentially economically available for research and management to laboratories with a PC and proper software. A shortcoming of these data sets is that SST is in 0.5°C intervals resulting in a loss of thermal resolution where differences in SST are more subtle. Another problem is that the images cannot be examined for cloud cover prior to downloading, however, the shortcomings are outweighed by the positive aspects of this service.

Processing satellite imagery can now be accomplished with several types of computers. The cost of a minicomputer, with an attached image processing system and appropriate software imagery, can vary up to about $200,000. A minicomputer allows much flexibility and speed. ELAS, a public domain software package developed by the U.S. government specifically for processing satellite imagery, is flexible, though not very user-friendly, and is commonly used in the United States. The flexibility of such a system and the many available shades of gray and colors allow the user to interpret an image in many ways. After processing, interpreting, color coding, and storing the image, a communications program and internal modem allow outside parties to dial the computer on any common carrier and access the image for downloading to a remote terminal.

Using a microcomputer is an alternative method for processing satellite imagery. Recent advances in PC technology have spurred the advent of several private companies which market software specifically written to process satellite imagery on a PC. Microcomputers and software are relatively inexpensive (e.g., about $30,000–$40,000 for a complete system). NMFS is now researching this technology for providing satellite images to vessels at sea, and software is being developed to process images for fisheries management and research applications.

Several shortcomings become apparent when processing satellite images with a PC. The speed of processing is relatively slow compared to minicomputers. This can be a severe hindrance when speed at providing an image is essential. Extremely large satellite scenes slow the process even further. Although a large range of colors and gray shades can be attained on a PC, the cost of the system increases accordingly. A large range of colors, especially gray shades, are necessary for proper interpretation of satellite images.

The most promising method of receiving satellite imagery at sea is by use of a PC equipped with proper software. A PC with a VGA board, color monitor, 10 Mb hard disk, and internal modem is a suitable system. In the GOM, a private company provides an offshore cellular service capable of sending images to vessels within about 150 miles of shore from Brownsville, Texas, to Panama City, Florida. Units to link the onboard system with the offshore cellular service cost about $3,000, and the cost of sending a single runs about $1.20 per minute. A recent test performed by NMFS scientists used just under 5 minutes to transmit a satellite SST image of the northeastern GOM to a research vessel at sea equipped with a PC and cellular telephone.

The previous method excludes those vessels in areas where an offshore cellular service is unavailable. In this case, the onboard PC must be equipped with a unidirectional tracking antenna capable of receiving signals from the INMARSAT communications satellite. Such a vessel would be able to receive imagery anywhere on the globe serviced by the INMARSAT satellites. The shortcomings are that the unidirectional receiving antenna and receiver are expensive, and the cost of transmission is higher. The antenna and receiver costs, at a minimum, approximately \$21,000, and transmission costs run approximately \$10 per minute. This would increase the cost of a 5-minute transmission from about \$6 to about \$50. This type of system would be limited to larger vessels.

A private corporation in Washington, D.C., routes Atlantic Ocean ship-to-shore telephone calls through INMARSAT to many European countries such as Spain and Denmark. A similar service is available for the Pacific Ocean. This suggests that images can be transmitted to vessels near any of the aforementioned countries. Using these communications satellite services to transmit images is expensive; however, this is the only other method available if an offshore cellular service is unavailable.

The basic need is to provide a georegistered satellite image of SST, derived from the NOAA satellites, to vessels at sea. When a new color-sensing satellite is placed in orbit in 1992, color imagery should become available for transmitting to research vessels. A grid of latitudes and longitudes or, preferably for many vessels, Loran-C lines because many vessels use Loran-C for location purposes, can be overlayed and transmitted with an image. Other types of information include bathymetric contours, bottom types, potential snags, and expert interpretation of the images with locations depicted on the image where certain species of fish might be found.

In the last 3 years, a GOM butterfishing industry has developed, and satellite images have proved useful at predicting where concentrations of butterfish might be found. Initial attempts by NMFS at transferring satellite information to a ship consisted of relaying front locations to a ship via a radio voice transmission. This method proved useful, but the shortcomings consisted of confusion created with the ship's personnel over what the scientist was trying to depict. Verbal translation of a visual image proved inadequate.

Another method consisted of converting a pseudocolor thermal image, displayed on an image-processing CRT, to a gray-scale plot on paper. Some interpretive information was overlayed on the gray-scale plot, and the entire plot was faxed (fax = facsimile transmission) via a common carrier (a land telephone line in this case) to vessels equipped with a fax machine and cellular telephone. Transmission was achieved by sending the signal through a commercial offshore cellular telephone service. Gray-scale plots, derived from satellite SST images of the northeastern GOM, were routinely faxed to two properly equipped vessels, and the plots aided the ship captains in determining where to fish.

The fax machines did not differentiate well between different shades of gray, thereby reducing the usable shades to about four or five. Fax machines also do not necessarily precisely recreate gray shades or even one shade homogeneously. The faxed output often darkened a shade of gray in one area and lightened it in another, further masking important features. In addition, a fax tends to form lines in a gray-scale plot, causing a plot to appear "streaky." For these reasons, faxed gray-scale plots were of poor quality, and much information was lost.

The most successful method used consisted of sending a pseudocolor image directly from the laboratory computer to a microcomputer, linked with a cellular telephone, aboard a fishing vessel. For demonstration purposes, the image was sent using INMARSAT instead of an offshore cellular telephone service. False color was added to a preprocessed image to depict 0.5°C temperature intervals. The image was compressed, using standard archiving software, and loaded into a communications program subdirectory. The communications program was commanded to stand by for sending. A researcher aboard the ship at sea used identical communications software on the vessel's PC to dial and interface with the laboratory computer and download the satellite image. The image was displayed on the PC monitor at sea using software developed by NMFS programmers.

Several butterfish vessels were equipped in 1989 with a microcomputer with color graphics capability. NMFS developed display software, a cellular phone, and an offshore cellular telephone interface device that allowed satellite transmission of compressed images directly from SSC to the fishing boats. The project was designed to demonstrate the commercial feasibility of using a microcomputer–cellular telephone link to transmit timely satellite images and predicted potential butterfish areas to selected fishing vessels at sea. The software to interpret the satellite imagery is based on an artificial intelligence expert system. A prototype expert system has recently been tested and is now entering the modification and refinement stage for future implementation.

REFERENCES

1. Kidwell, K. B. 1985. *NOAA polar orbital data (TIROS-N, NOAA- 6, NOAA-7, NOAA-8, and NOAA-9) users guide.* NOAA/NESDIS, National Climatic Data Center, Satellite Data Services Division, Washington, D.C.
2. Gower, J. F. R. 1982. General overview of the nature and use of satellite remote sensing data for fisheries applications. NAFO Sci. Coun. Studies 4:7–19.
3. Putnam, E. S. (ed.). 1987. *System concept for wide-field-view observations of ocean phenomena from space.* Report of the Joint EOSAT/NASA SeaWiFS Working Group, National Aeronautics and Space Administration, Washington, D.C.
4. Gledhill, c., M. Grace, R. C. Herron, A. J. Kemmerer, T. D. Leming, L. Regier, C. Roithmayr, W. Seidel, and J. Watson. 1988. *1988 annual report: latent resources (butterfish, coastal herrings and associated species)* (R. C. Herron, ed). National Marine Fisheries Service, Southeast Fisheries Center, Mississippi Laboratories, Pascagoula, Miss.
5. Steen, J. P., Jr., R. C. Herron, and C. E. Eleuterius. 1988. The description of a cold core ring formation on the Mississippi-Alabama-Florida continental shelf during the winter of 1988. Paper presented at the Mineral Management Service's Information Transfer Meeting, October, 1988, New Orleans.
6. Leming, T. D. 1981. *Interim report—FY 81: October 1, 1980—March 31, 1981.* Unpublished NMFS Report, Task Number 88ASF4F0, National Marine Fisheries Service, Mississippi Laboratories, Pascagoula, Miss.
7. Maul, G. A., F. Williams, M. Roffer, and F. M. Sousa. 1984. Remotely sensed oceanographic pattern and variability of bluefin tuna catch in the Gulf of Mexico. Acta 7:469–479.

8. Leming, T. D., R. C. Herron, and X. Wang. 1987. Larval bluefin tuna (*Thunnus thynnus*) and mesoscale Gulf of Mexico Circulation. EOS 68(50):1751.

9. Richards, W. J., T. Leming, M. F. McGowan, J. T. Lamkin, and S. Kelly-Fraga. 1988. The distribution of fish larvae in relation to hydrographic features of the Loop Current boundary in the Gulf of Mexico. Rapp. P.–v. Reun. Cons. int. Explor. Mer 191:169–176.

10. Kemmerer, A. J. 1980. Environmental preferences and behavior patterns of Gulf menhaden (*Brevoortis patronus*) inferred from fishing and remotely sensed data, in *Fish Behavior and Its Use in the Capture and Culture of Fishes* (Bardach, J. E., J. J. Magnuson, R. C. May, and J. M. Reinhart, ed.), ICLARM Conference Proceedings 5, International Center for Living Aquatic Resources Management, Manila, Philippines, pp. 345–370.

11. Vanselous, T. M., A. J. Kemmerer, W. A. Hovis, and D. K. Clark. 1978. Marine applications of the NIMBUS–G Coastal Zone Color Scanner. National Marine Fisheries Service, Southeast Fisheries Center, Stennis Space Center, Miss. SEFC Contribution No. 78–24F.

12. Herron, R. C. 1986. Utilizing satellite derived data and discriminant function analysis to predict areas of Gulf menhaden (*Brevoortia patronus*) abundance in the northern Gulf of Mexico. Internal Report, National Marine Fisheries Service, Southeast Fisheries Center, Mississippi Laboratories, Stennis Space Center, Miss.

13. Stumpf, R. P. 1987. Detection of turbid water in sunglint regions of remote sensing imagery. Internal Report, National Oceanic and Atmospheric Administration, Assessment and Information Services Center, Washington, D.C.

14. Grimes, C. B., J. H. Finucane, J. T. Brucks, and R. C. Herron. 1987. Spatial distribution and abundance of ichthyoplankton about the Mississippi River plume: The possible importance of the plume to recruitment success of economically important species. paper presented at the Es-

tuarine Research Federation's Annual Meeting, October 1987, New Orleans.

15. Grimes, C. B., J. J. Govoni, D. E. Hoss, J. H. Finucane, D. A. DeVries, K. L. Lang, D. B. White, J. T. Brucks, R. C. Herron, A. B. Powell, and S. J. Warlen. 1988. Aggregation of ichthyoplankton at the Mississippi river plume front: Possible role of the plume in recruitment success. Paper presented at the American Association for the Advancement of Science's Annual Meeting, February 1988, Boston.

16. Grimes, C. B., J. H. Finucane, V. M. Pettigrew, J. T. Brucks and R. C. Herron. 1988. King mackerel early life stages in the Gulf of Mexico: Possible association with the Mississippi River Plume. Paper presented at the American Fisheries Society's Florida Chapter's Annual Meeting, February, 1988, Brooksville.

17. Dagg, M. J., P. B. Ortner, and F. Al-Yamaru. 1988. Wintertime distribution and abundance of copepod nauplii in the northern Gulf of Mexico. *Fisheries Bulletin U. S.* 86(2): 319–330.

18. Govoni, J. J., D. E. Hoss, and A. J. Chester. 1984. The interaction of larval Gulf menhaden, *Brevoortia patronus*, and spot, *Leiostomus xanthurus*, with the Mississippi River plume. EOS 64:1022.

19. Olson, D. B., and R. H. Backus. 1985. The concentrations of organisms at fronts: A cold-water fish and a warm-core Gulf Stream ring. J. Marine Research 43:113–137.

20. Richardson, K. 1985. Plankton distribution and activity in the North Sea/Skagerrak-Kattegat frontal area in April 1984. Marine Ecology Progress Series 26:233–244.

21. Richardson, K., M. R. Heath, and S. M. Pedersen. 1986. Studies of a larval herring (*Clupea Harengus* L.) patch in the Buchan area. III. Phytoplankton distribution and primary productivity in relation to hydrographic features. Dana 6:25–26.

22. Leming, T. D., and R. C. Herron. 1986. Associations of large schools of butterfish (*Peprilus burti*) with thermal fronts. EOS 67(44):981.

23. Herron, R. C., T. D. Leming, and J. Li. 1987. Gulf butterfish (*Peprilus burti*) distribution in the northeastern Gulf of Mexico and environmental relationships. EOS 68(50):1724.

24. Leming, T. D. 1987. BUFERSS: An expert system for interpreting satellite oceanographic imagery as a guide to butterfish distributions in the northern Gulf of Mexico. Internal Report, National Marine Fisheries Service, Southeast Fisheries Center, Mississippi Laboratories, Pascagoula, Miss.

25. Schowengerdt, R. A. 1983. *Techniques for image processing and classification in remote sensing*, Academic Press, Orlando, Fla.

26. Herron, R. C., T. D. Leming, and J. Li. 1989. Satellite detected fronts and butterfish aggregations in the northeastern Gulf of Mexico. Continental Shelf Research 9(6):569–588.

V

Computer Vision Technology for the Fish Industry

10

Image Analysis in the Fish and Food Industries

NORVAL J. C. STRACHAN and C. K. MURRAY *Ministry of Agriculture, Fisheries & Food, Aberdeen, Scotland*

1. Introduction
2. Present and Potential Uses of Image Analysis
3. Image Analysis at Torry Research Station
   3.1 Species and Size Grading
   3.2 Morphological and Particle Analysis
4. Summary
   Appendix 1
   Appendix 2
   References

1. INTRODUCTION

Torry Research Station is the U.K. government laboratory concerned with the investigation of all aspects of fish technology. In order to keep up with the advances of microelectronics, a working party was set up in 1986 to report on the needs of the industry in respect of electronic applications and to identify areas in which Torry should work. One of the main conclusions of the working party was that research and development should be done on image analysis and pattern recognition for grading and quality control of fish and fish products. It was felt that in using these techniques, some of the repetitive, menial, labor-intensive tasks could be automated, resulting in cheaper products of higher yield and quality.

There are many books and journals available on the subject of image analysis (Appendices 1 and 2). These are not specific to the fish and food industries. However, they state and explain the current advances, theories, and ideas of image analysis, and applications to problems in the food industry can be based on these concepts.

A basic image analysis system is outlined in Figure 1. The illumination of the object can be from a source emitting any part of the electromagnetic spectrum (i.e., visible, ultraviolet, infrared, and X-ray) and from any direction. The image sensor is chosen according to what parameters are required to be sensed. For example, when using visible light, a CCD (charge coupled device) or a vidicon tube camera may be used.

The analog signal from the sensor is then digitized using an A/D converter. The digitized image can have various array sizes, which are usually no bigger than 512 × 512 pixels. The amount of bit planes per pixel can be chosen from one (binary image) to normally no greater than 12. If true color is required, then each

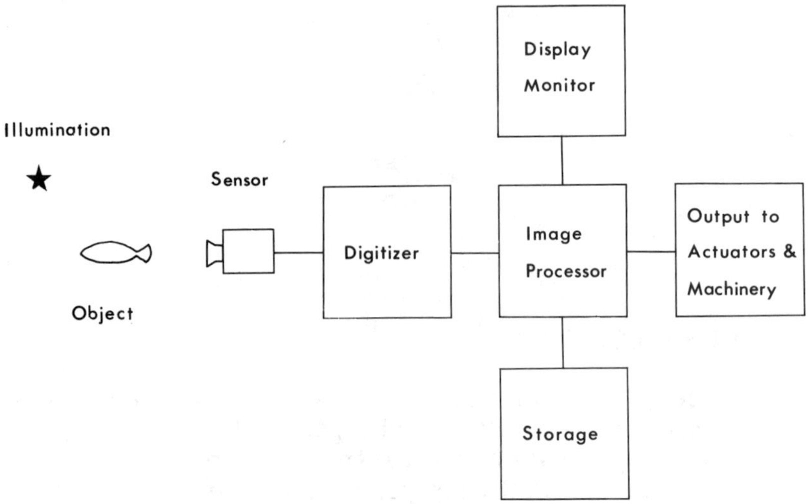

Figure 1 Elements of an image analysis system.

pixel of the image must be digitized to give a red, a green, and a blue component. The number of bit planes chosen for each color again is usually at most 12. The digitized images are then processed by a computer, which may be conventional serial one (e.g., IBM PC (1) or SUN (2) workstation) or a specialized parallel one (e.g., INMOS TRANSPUTER (3)), depending on the type and speed of processing required.

The output from the image processor can then be used to operate actuators and machinery, e.g., air and water jets, gates, and robot arms.

2. PRESENT AND POTENTIAL USES OF IMAGE ANALYSIS

Image analysis is being used and researched in the fishing industry to sort herring by sex, to detect bones in fish fillets, to detect parasites, to measure the size of the fish to control a gutting machine, and to sort fish by species, size, and weight.

Neptune Dynamics Ltd. (4) manufactures the HD1S Herring Sex Discriminator. The machine separates mature fresh herring by sex. A source of variable intensity infrared light is directed at the belly of the fish where the eggs or milt are. The amount of energy transmitted through the fish is sensed, electronically analyzed, and the sex of the fish is determined by a simple thresholding technique.

Lumetech (5) produces a machine that can reduce bones and fins to 10–20% of the usual incidence in fish fillets. The system works by illuminating the fillets with ultraviolet light, which causes the bones to fluoresce. The bones are then detected by a thresholding technique using water jets. One problem with this technique is that it is basically a surface one as explained by Huss et al. (6), and bones that are inside the fillet may not be detected.

Brataas (7) states that at least 8–10 different institutes or companies are trying to develop a machine to detect parasites in fish. Different radiations are being used (ultraviolet (8), visible, infrared, X-ray, ultrasound, laser, and scanning laser acoustic microscope (SLAM)). No commercial detection device has yet been developed.

Halvorsen (9) has described a machine which uses a vertical array of LEDs and a corresponding array of photodetectors to detect the image of a fish. The image is then processed using a computer that gives outputs to a gutting machine to give each individual fish optimal treatment according to its size and weight.

There have been various attempts to sort fish by species and size using image analysis. A machine has been developed by Tayama et al. (10) to identify fish by species and sort by size using an optical sensor to input binary pictures, a computer, and image recognition techniques. The dimensions of various parts of the body are ascertained and then, by calculating characteristic shape descriptors such as thickness of the tail stalk and body length ratios (32 descriptors are used), a species identification formula is produced for four different species of fish, which sorts with a quoted reliability of 95%. The system has two channels and has a throughput of 200 fish per minute. Wagner and Walter (11) suggest using the back line of a fish (from the nose to the tail stalk) as a criterion for distinguishing between the different species of fish. Wagner et al. (12) then goes on to describe a similar method to that of Tayama where seven shape descriptors are measured to sort nine species of fish with a reliability of 90%.

fisheries Technology Group Ltd., Canada (13), claim to have made a machine that can sort fish by species and size with a reliability of 99%. The backlit fish pass under a conventional video camera, and the image is sent to a computer where it is analyzed using a fish species recognition formula (no details and no experimental data are given).

In the food industry, image analysis machines are being developed and used for a large number of purposes, some of which are outlined below.

Glafascan (14) is a machine developed by the AFRC Food Research Institute at Bristol by Newman (15, 16), which uses image analysis to detect the fat/lean ratio of boneless fresh and frozen meat as it passes along a conveyer belt. The Food Research Institute are doing work with Bristol University (17) to perform fundamental research into the design and use of advanced computer vision and intelligent, sensor-guided robots for meat grading, cutting, handling, packaging, and cooking.

Loctronic Graders Ltd. (18), in conjunction with the Essex Electronics Centre (19), are currently developing a real time true-color sorting system using image analysis for inspection of fruit, vegetables, and other food products.

Davies (20) describes methods for the rapid inspection of circular products, in particular biscuits. These techniques include an inherently rapid procedure for the precise location of the centres of circular products and radial intensity histograms to aid detailed surface scrutiny. This approach was found to be capable of detecting and identifying correctly a variety of faults, including broken and overlapping biscuits, products having only a partial coating of chocolate or jam, surface blemishes, and protuberances on their edges.

Sarker and Wolfe (21) have developed and implemented image analysis and pattern recognition algorithms for orientation of fresh market tomatoes and classification has an error rate of less than 8%. It is concluded that computer vision is effective at sorting tomatoes and that it approaches and in some cases exceeds the capabilities of human inspectors.

Devalaeminck (22) describes a robot vision system engineered by High Technology Holland, which provides a fully automatic mushroom-harvesting system. Two solid state cameras are used and from their image outputs a computer determines the position, height, and size of the mushrooms. The computer then decides which mushroom to pick first, and it does this by controlling a robot arm. The system does not function as reliably and as quickly as desired, and further work is being done to improve this.

Schatzki et al. (23) describe an X-ray method using image analysis to detect agricultural contraband (e.g., fruit such as grapes) in passenger luggage. It is suggested that most contraband can be distinguished from other material by an elliptic, rather than rectangular, cross section. An algorithm is described which recognizes the cross section by eroding the absolute gradient of the image. It is shown that the algorithm detects desired items as small as 1–2 cm.

Throughout the fish and food industries, image analysis systems, some of which have been described here, are being developed and designed. Most of these systems use binary images only, thus cutting down on the amount of processing power required and reducing the cost of the equipment.

3. IMAGE ANALYSIS AT TORRY RESEARCH STATION

At Torry there are two image analysis systems. Both systems use conventional video cameras and Imaging Technology Inc. (24) digitizing and image processing boards, which can give 12-bit resolution and true color outputs. The host computers are an IBM PC–compatible and a Sun 3/160C workstation, respectively.

Applications of Image Analysis on fish and fish products is being studied at Torry in two areas: 1) species and size grading, and 2) morphological and particle analysis.

3.1 Species and Size Grading

Work is being carried out on the sorting of fish by species and size.

A general purpose pattern recognition machine is CRS WISARD manufactured by CRS Computer Recognition Systems (25). WISARD is an acronym for Wilkie, Stonham, and Alexanders (the inventors) Recognition Device. It has been used in col-

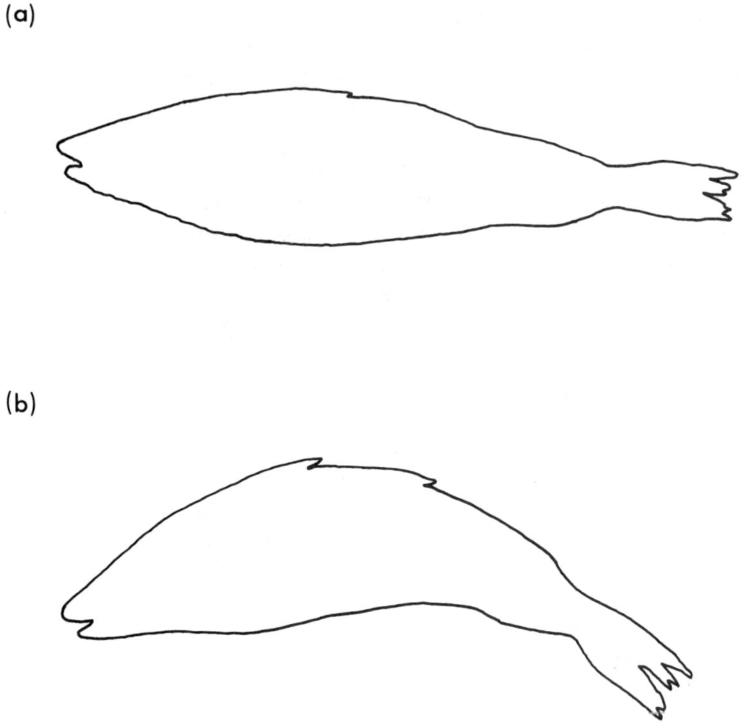

Figure 2 A fish (a) before and (b) after being flexed.

laboration with Brunel University (26) to sort fish by species. WISARD stores patterns, and when a new pattern is presented to it can give an estimate of how similar the new pattern is when compared with the learned ones. The machine has a special purpose semi-parallel architecture and is therefore very fast and can process television resolution images at video frame rate of one picture every 40 milliseconds. The results of sorting fish by species were quite promising using this machine, but there were a number of drawbacks. There was no normalization of the image for size, rotation, and position, nor would the system allow for the flexible nature of fish (Fig. 2). Thus it was decided to stop work with this machine for the time being and concentrate on other techniques.

Three methods of sorting seven fish species (Fig. 3) by shape analysis of images have been studied by Strachan et al. (27). The first method uses the invariant moments technique described by Hu (28). The technique is not computationally intensive. Six invariant moments were used to describe the fish shapes, and the sorting reliability of the method was 73.3%. A disadvantage of the moments is that they had no obvious physical interpretation. The second method compares a sample fish shape with a reference fish shape (a template). The sample shape is superimposed by a

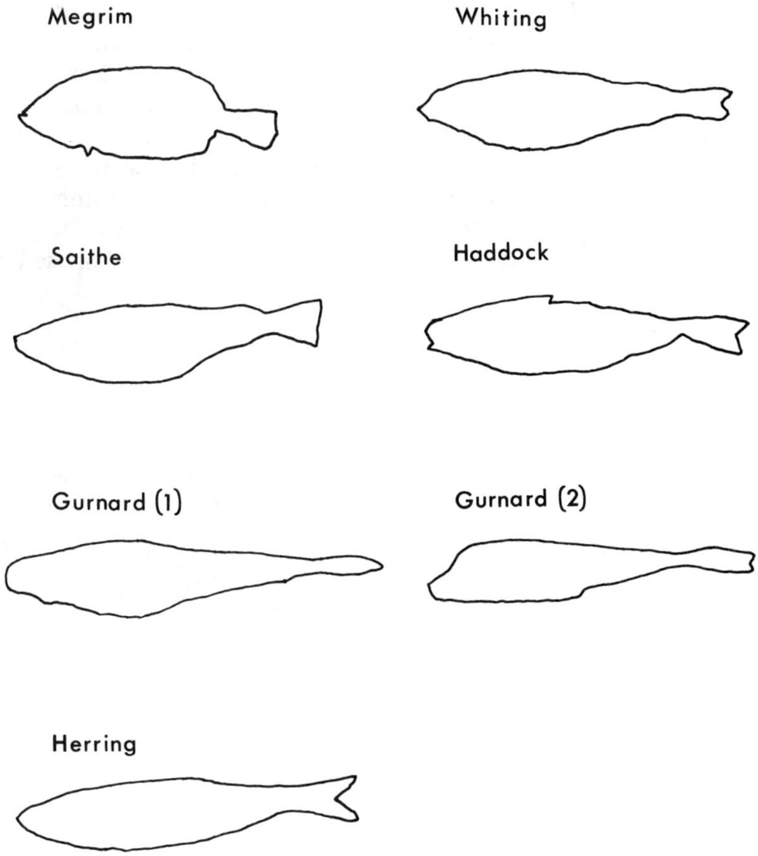

Figure 3 Computer diagrams of the seven species of fish used for analysis.

sequence of translations, rotations, and scaling of the template using an optimization technique until the mismatch factor (a figure of how well the sample matches the template) is a minimum. The minimum value of mismatch is then used to compare the different species of fish. The method is computationally very intensive, requiring 1 1/2 hours to compare each pair of fish on the Prime 9650 minicomputer. The sorting reliability was poor—63.3%.

The third method uses eleven shape descriptors, as shown in Fig. 4, and is similar to that of Tayama and Wagner. This method proves to be the most reliable (90%). The descriptors being used all had physical meaning, and the method was computationally unintensive. If higher resolution images and more shape descriptors are used, reliability may be improved.

At present all of these methods are unable to cope with fish that have been flexed or distorted. The main problem is that the geometric center of area and the principal axis of the fish used to position the fish within the image are not good references if the fish is flexed. Work is now being carried out to find an alternative reference axis to alleviate this problem. Once this problem has been solved, grey scale as well as color analysis of the fish will be studied.

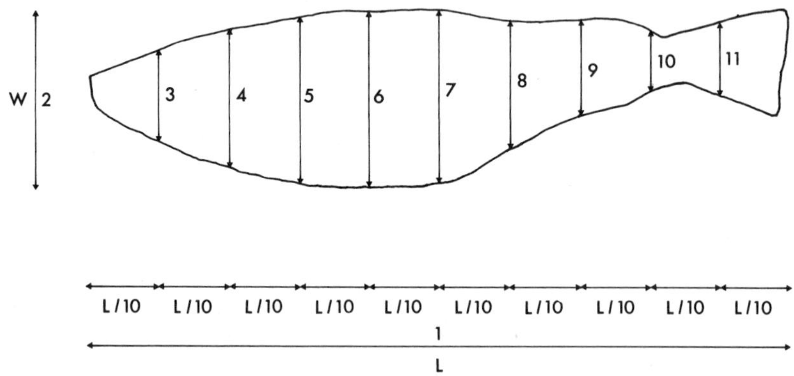

Figure 4 A fish shape with the 11 shape descriptors marked on it.

3.2 Morphological and Particle Analysis

The eating quality of foods depends not only on their flavor and visual appearance but also on their textural properties. This feature is particularly important in the manufacture of new and analogue products. Products can now be "engineered" by adding retexturing agents. The purpose of this work is to correlate the quantity and spatial distribution of these agents with mouthfeel, hence improving existing products and developing new ones.

It is unknown what particle sizes and shape structure such as fibers the mouth can detect, and these are being investigated in the range of 1 μm to 1 mm. Slices of analog foods prepared at Torry are taken and stained to highlight globules, e.g., of fat and starch. Images of these are then taken using a video camera and a light microscope. Eventually a scanning electron microscope will also be used. These images are then analyzed by computer to calculate the distribution and the quantity of the stained elements. A relationship between this data and the results from a taste panel will be established. This will now be used to develop products.

4. SUMMARY

Image analysis systems have and are increasingly being designed for use in the fish and food industries. Only simple thresholding machines have passed the prototype stage and are being used in factories. More sophisticated machines that use color, grey scale, and/or robot manipulators are being developed and will be implemented in the next 10 years. Torry Research Station has invested in the latest equipment in this field to do fundamental research and to keep at the forefront of this technology.

APPENDIX 1

Image Analysis Reference Books

Rosenfield, A., and A. C. Kak. 1982. *Digital Picture Processing*, Vols 1 and 2. Academic Press Inc, New York.

Ballard, D. H., and C. M. Brown. 1982. *Computer Vision*. Prentice-Hall Inc., Englewood Cliffs, N.J.

Pavlidis, T. 1977. *Structural Pattern Recognition.* Springer-Verlag, Berlin, Heidelberg, New York.

Pavlidis, T. 1982. *Algorithms for Graphics and Image Processing.* Springer-Verlag, Berlin, Heidelberg, Germany.

Andrews, H. C. 1970. *Computer Techniques in Image Processing.* Academic Press, New York.

Castleman, K. R. 1979. *Digital Image Processing.* Prentice-Hall, Englewood Cliffs, N.J.

Duda, R. O., and P. E. Hart. 1973. *Pattern Classification and Scene Analysis.* Wiley, New York.

Hall, E. L. 1979. *Computer Image Processing and Recognition.* Academic Press, New York.

Pratt, W. K. 1978. *Digital Image Processing.* Wiley, New York.

APPENDIX 2

Image Analysis Journals

Pattern Recognition. Pergamon Press Ltd, U.K..

Pattern Recognition Letters. Elsevier Science Publishers B.U., Amsterdam, Netherlands.

Computer Vision Graphics and Image Processing. Academic Press Inc., New York.

IEEE Transactions on Pattern Recognition and Machine Intelligence. *IEEE, New York.*

Visual Computer. Springer-Verlag, New York.

REFERENCES

1. IBM UK Ltd., PO Box 41, North Harbour, Portsmouth, PO6 3AU, England.
2. Sun Microsystem UK Ltd., 38 Melville Street, Edinburgh, EH3 7TW. Scotland.

3. Inmos Transputer, 1000 Aztec West, Almondsbury, Bristol, BS12 4SQ, England.

4. Neptune Dynamics Limited, 1768 East Hastings St, Vancouver, British Columbia, Canada, V5L 1S9.

5. Lumetech, Carlsberg Research Center, G1. Carlsberg Vej 10, DK–2500 Valby, Denmark.

6. Huss, H. H., P. Sigsgaard, and S. A. Jensen. 1985. Fluorescence of fish bones. J. of Food Protection 48(5):393–396.

7. Brataas, R. 1988. *Detection of Parasites in Cod* WEFTA Tromsø, Norway.

8. Pippy, J. H. C. 1970. Use of ultraviolet light to find parasitic nematodes in situ. J. Fish. Res. Bd. Canada 27:963–965.

9. Halvorsen, S. 1984. Microprocessor control in fish processing equipment. in *Engineering and Food*, Vol. 2, *Processing Applications* (McKenzie, ed.)

10. Tayama, I., M. Shimdate, N. Kubota, and Y. Normura. 1982. Application for optical sensor to fish sorting, Reito (Tokyo), Refrigeration 57(661):1146–1150.

11. Wagner, H., and J.H. Walter. 1983. Sorting of species of marine fish. 1. Statistical geometric definition of species of fish. Lebensmittelindustrie 30(8):375–376.

12. Wagner, H., U. Schmidt, and J. H. Rudek. 1987. Distinction between species of sea fish. Lebensmittelindustrie 34: 20–23.

13. Fisheries Technology Group Ltd, FMS 1000, PO Box 9190, St Johns, Newfoundland, A1A 2X9, Canada.

14. Glafascan Ltd, Partridge Green, Horsham, West Sussex, RH13 8JP.

15. Newman, P. B. 1984. The use of video image analysis for quantitative measurement of visible fat and lean in meat: Part 1—Boneless fresh and cured meat. Meat Science 10:87–100.

16. Newman, P. B. 1984. The use of video image analysis for quantitative measurement of fatness in meat: Part 2—Comparison of VIA, visual assessment and total chemical fat estimation in a commercial environment. Meat Science 10:161–166.

17. Khodabandehloo, K., C. Bailey, and P. B. Newman. 1987. *Vision Technology and Robotics in the Food Industry*. Institute of Mechanical Engineers, Food Manufacture using Robotics and Flexible Automation. London, England.

18. Loctronic Graders Ltd, Danbury, Chelmsford, Essex, CM3 4NH, England.

19. Essex Electronics Centre, University of Essex, Wivenhoe Park, Colchester, Essex, England.

20. Davies, E. R. 1984. Design of cost effective systems for the inspection of certain food products during manufacture. *Proc. 4th Conf. on Robot Vision and Sensory Controls*, London, 9–11 Oct. pp. 437–446.

21. Sarkar, N., and R. R. Wolfe. 1985. Feature extraction techniques for sorting tomatoes by computer vision. Transactions of the American Society of Agricultural Engineers, pp. 970–979.

22. Devalaeminck, R. M. 1985. *Vision Systems and Robotics in Food Processing*. Agrimation: Conf. Proc., Chicago, 1985, pp. 27–37.

23. Schatzki, T. F., A. Grossman, and R. Young. 1983. Recognition of agricultural objects by shape. IEEE Transactions on Pattern Analysis & Machine Intelligence, PAMI–5, No 6. pp. 645–653.

24. Imaging Technology Inc., 600 West Cummings Park, Woburn, MA 02801, USA.

25. Computer Recognition Systems Ltd., The Business Centre, Molly Millars Lane, Wokingham, Berks, England.

26. Brunel Univ., Dept of Electrical Engineering and Cybeneretics Division, Uxbridge, Middlesex, UB8 3PH.

27. Strachan, N. J. C., P. Nesvadba, and A. R. Allen. 1990. Fish Species Recognition by Shape Analysis of Images. Pattern Recognition, Vol. 23, No. 5, pp. 539–544.

28. Hu, M. K. 1962. Visual pattern recognition by moment invariants. IRE Transactions on Information Theory, Vol IT–8, pp. 179–187.

11

Laboratory Work Using the DT 2851 Frame Grabber: Application of Computer Vision to Inspection of Fish

ERIC BARDIN *Technical University of Denmark, Lyngby, Denmark*

♦♦

1. Introduction
2. Experimental Setup
3. Preparing the Practical Work
4. Training with Binarized Images
5. Isolating a Fish Fillet in an Image
    5.1 Acquiring Correct Images
    5.2 Thresholding, Slicing
6. Finding a Surface Defect in a Fillet (blood spot, residual skin)
    6.1 With a White Background
    6.2 With a Reddish-Brown Background

7. Introduction of Filters in the Illumination and Sensing Systems
8. Window to Detect Defects
9. Extraction of Edges
   9.1 Using Highpass and Lowpass Filters
   9.2 Using Laplacian Filter
   Appendix
   Reference

♦♦

1. INTRODUCTION

After reading this chapter, you should have acquired basic notions about: the different parts of a vision system (illumination, sensing, and computing systems) and the way to manage them optimally; some basic image analysis techniques such as grabbing of a picture, histogramming, thresholding, slicing, windowing, and filtering operations; and the possible applications of vision techniques to fish or fish fillet inspection and the problem met in these applications.

2. EXPERIMENTAL SETUP

In order to do the laboratory work, you must assemble the following materials:

1. PC with the DT 2851 frame grabber installed (with a Vaxmate keyboard)
2. DT 2851 grey connection wire bundle (eight plugs at each end)
3. 1 multiple power plug
4. 1 RGB video monitor (with three red, green, and blue input lines)
5. 1 CCD area camera with lens filters and 1 power supply transformer
6. 1 physical support for the camera and the illumination filters (see Figure 1 for disposition, two 60-cm-long wood lathes could be useful to hold the filters)
7. 1 or 2 bolts in order to fasten the camera on its support
8. 1 umbrella-like reflector painted white inside
9. 1 125 W HQL DE LUXE OSRAM bulb and the special power transformer for it

10. 1 desk lamp with white machine paper leaves (as light diffuser)
11. blue, red, and green plexiglas slices (as filters for illumination with the desk lamp)
12. 1 wide wood or plastic sheet or blanket (needed to protect the scenes recorded from daylight illumination when laboratory work is performed during the day)
13. 4 2-m-long coaxial cables
14. large pieces of tissue (used as background) of several colors + white
15. 1 large transparent plastic sheet or bag to cover the background tissue and to protect the latter from the fish fillet
16. fish fillets (at least 1 normal, 1 blood spotted, 1 with skin residue)
17. 1 calculator (+,–,* and / operations)
18. the Iristutor User's Manual
19. the document "User Guidelines for DT 2851 Frame Grabber and IRIS Tutor" (1)

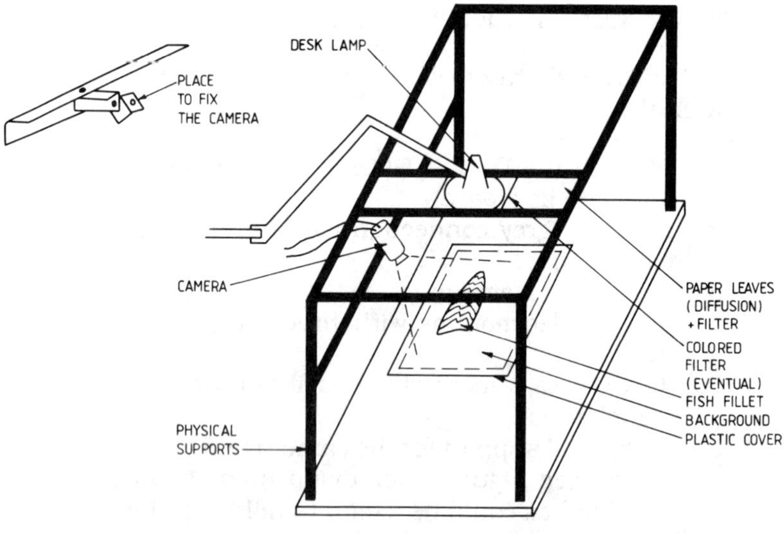

Figure 1 Disposition of the lighting and sensing systems.

3. PREPARING THE PRACTICAL WORK

Before starting the work with IRIS Tutor, you first have to install the materials collected:

Install camera and lighting system, using the necessary physical supports as described in Figure 1; at the beginning, however, install the OSRAM lamp and the umbrella-like reflector instead of the desk lamp (at the same location).

Do not forget to protect the objects you scan from the day or room light with the sheet or the blanket if necessary.

Perform the connections described in paragraph 2 of the "User Guidelines for DT2851 and Iristutor."

Connect power to all devices.

Switch on the PC and its monitor, the video monitor and the camera power transformer (if necessary).

Once the PC initialization is achieved, the [c:/] prompt is displayed.
Enter the IRIS subdirectory by typing:

CD IRIS

(plus return, *return is to be typed after every command*). The whole IRIS Tutor software package is implemented in this directory.

You can read a macro by typing:

TYPE <file_name> (plus return...)

where file_name is the name of the file containing the macro you want to modify.

You can invoke the Quick C editor from this directory, in order to create or modify macro files, by typing:

QC <file_name> (plus return...)

where file_name is the name of the file containing the macro you want to modify.

To invoke the type command or the Quick C editor, remember you must exit Iris Tutor (by typing EXIT) if you use it.

To enter Iris Tutor, type:

IR

4. TRAINING WITH BINARIZED IMAGES

To get acquainted with some simple commands and notions, we are now going to use a special purpose macro called TRAINING.TP . The image supplied by the camera does not need to be a special image for this *training*. The only requirement is a certain variety and contrast in the image.

Note: A binarized image is one whose pixels all have a value of 0 or 255.

Execute the macro by typing:

EXECUTE TRAINING.TP

and read comments while the macro executes; use Ref. 1 or IRIS Tutor manual to get details about a command and see more precisely what tasks the different commands are performing.

Executc the INITIAL.TP MACRO (initialization of the frame grabber) by typing:

EXECUTE INITIAL.TP

Remember the basic commands with these examples:

ACQUIRE 1 FRAME TO BUFFER 1 grabs an image in buffer 1.

HISTOGRAM BUFFER 0 displays the histogram of the image in buffer 0.

5. ISOLATING A FISH FILLET IN AN IMAGE

Take the white background, place the plastic cover on it, and a normal fish fillet on the cover.

5.1 Acquiring Correct Images

Changing eventually camera lens aperture and focus, display an image on the video monitor satisfying these conditions:

The fish fillet is entirely in the image, and the background occupies the whole remaining room.

The image is crisp.

There is no reflected light visible in the image (if the light is vertically falling onto the object surface, the angle of the camera main axis with a vertical line should be at least 25 degrees).

The image is unsaturated (i.e., the right histogram value (255-pixel value) must not be a maximum); otherwise there would be a loss of contrast in the image; this is to be checked with the histogram function once the image has been acquired. Be also aware that the histogram has to be as wide as possible: This is another condition for a good contrast in the image (if for instance all pixel values are under 128, it will be more difficult to differentiate the object scanned from the background).

When all these conditions seem to be verified, execute the RECORD.TP macro that will acquire the images.

Check with the histogram command that your image is unsaturated. If not, repeat section 5.1. Note that the first command in the RECORD. TP macro is 'DISPLAY CAMERA', so that you can see what images you will acquire.

5.2 Thresholding, Slicing

The histogram should normally look like that shown in Figure 2. Isolating an object from the background in an image necessitates a binarization operation of the image.

One way to binarize an image is to threshold it, which is done in two steps:

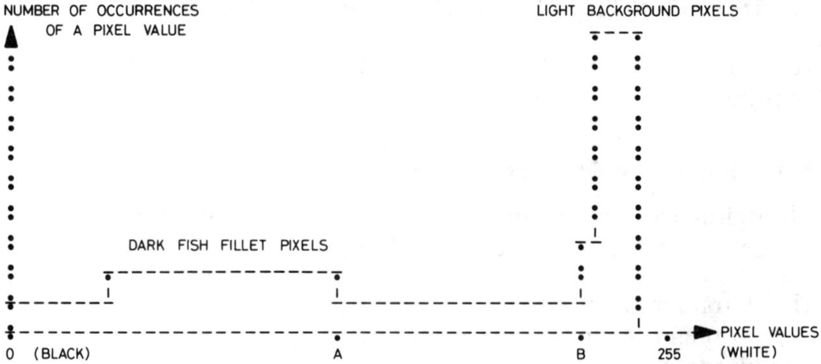

Figure 2 Histogram of a normal fish fillet image (white background).

1. Subtract the threshold from all pixel values in the image (movement of the histogram towards the left).
2. Multiply the result of step 1 by the value 255 (maximum value) to saturate pixels whose value is not null.

The result of a thresholding operation is to differentiate pixels with higher and lower values than a threshold in an image.

Another way of binarizing an image is to slice it; this binarization selects an interval of pixel values in the histogram, and is done in two steps:

1. Thresholding of the same image in two different buffers, with two different thresholds that delimit the interval to isolate.
2. XOR operation between both buffers.

Let us detail this slicing operation, and assume you have already recorded the image in both buffers 1 and 0; now, perform two thresholding operations (using two threshold values such as A and B in Figure 2) in order to isolate the fillet in the image:

Add to *buffer 0* the *negative* A threshold value (do not use the ADD command but OFFSET); then order a histogram to see the variation.

Repeat the same operation with *buffer 1* using the negative B threshold value.

Execute the SLICING.TP macro which does the following:

ends the thresholding operations

displays the thresholded images in buffers 0 and 1; the fish fillet is more or less well isolated in both cases

displays a binarized image. This image is the result of the slicing operation between A and B pixel values (XOR operation on the two previous images); it shows the differences between both thresholded images, and consequently how important the choice of the threshold is in order to isolate an object from the background; if the background hue is very homogeneous, the best threshold choice is near 3.

6. FINDING A SURFACE DEFECT IN A FILLET (BLOOD SPOT, RESIDUAL SKIN)

6.1 White Background

Go on using the same background, plastic cover, illumination and sensing system (no filters); put a fish fillet with blood spots under the camera field of view.

Execute the INITIAL.TP macro.

Display an image which satisfies the conditions described in Section 5.1. and in which the defect is to be seen easily.

When all these conditions are satisfied, execute the RECORD.TP macro. If the histogram is not good enough (good means wide histogram, unsaturated image), execute this macro again. The optimum result you should obtain is shown in Figure 3.

To isolate the defect, you have to threshold the image. Choose a threshold like C in Figure 3.

Before beginning the thresholding operation, you can verify the validity of your threshold choice, or find this threshold value, in either of two ways:

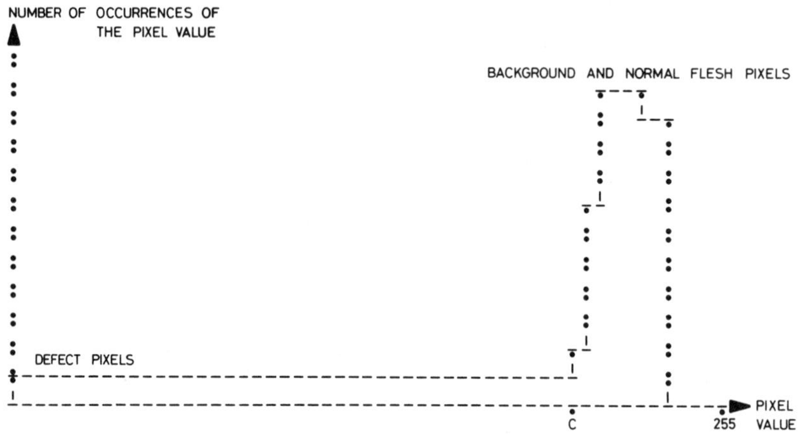

Figure 3 Histogram of an image of a fillet with blood spots (white background).

Execute the FINDWIND.TP macro, and move the cursor into the defect region to know the real pixel values of the defect.

Modification of the output table 0 in order to display in a particular color the pixels with a value under the threshold; this operation gives you information about the pixels which will be isolated by the chosen threshold (some are not part of the defect).

Now, threshold the image in buffer 0:

First, offset your image by the negative C value.

Then run the THRES.TP macro; at the end of this execution, the buffer 0, which is displayed, should contain the defect in white on a black background.

If the result is not good enough, you can repeat from the start of this paragraph.

You can then estimate the defect area by summing all pixels in buffer 0, and dividing the result by 255 (use the small calculator for it).

If you have time, you can repeat all operations in Section 6.1 on a fillet with skin residue. It is probably not necessary to change the lens aperture; but if it is needed, take note of the previous lens aperture in order to compare it with the new one: it will be possible to see the differences between sensor requirements for the detection of two different fillet defects.

6.2. Reddish-Brown Background

Take the reddish-brown background and the plastic cover and put the fillet with blood spots on the cover under the camera's field of view. Execute the INITIAL.TP macro.

Display an image which satisfies the conditions listed in Section 5.1. besides the following ones:

The defect is to be seen easily in the image.
The differences between the background, the normal flesh and the defect colors must be as wide as possible.

When all these conditions are met, execute the RECORD.TP macro. Try to obtain an histogram as in Figure 4 (execute the RECORD.TP macro as many times as necessary). In fact, using the reddish-brown tissue, the defect should appear lighter than the background, and of course darker than the fillet; but it is not obvious, and sometimes impossible, to obtain a defect peak.

You need to isolate the defect pixels, using a slicing operation. If your histogram looks like Figure 4, choose thresholds like

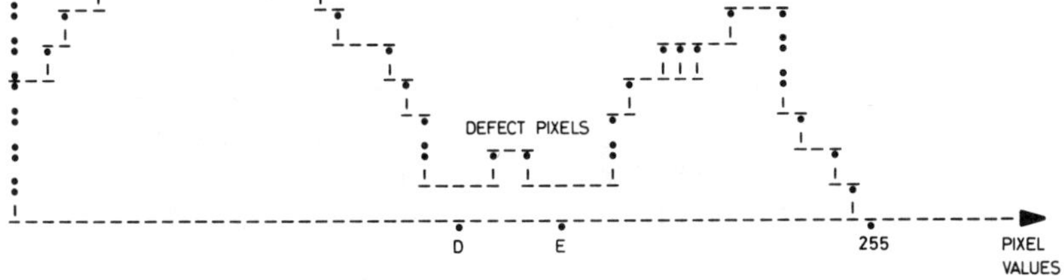

Figure 4 Histogram of a fillet image with defect (reddish-brown background).

D and E. If there is no peak in the histogram showing the defect pixels, execute the FINDWIND.TP macro and move the cursor in order to determine the values of D and E.

Before beginning the thresholding operations, you can verify the validity of your threshold choice in either of two ways:

Execute the FINDWIND.TP macro (if you have not already done that) and move the cursor inside the defect region to know the real pixel values of the defect.

Modify the output table 0 in order to display in a particular color the pixels with a value between the two thresholds; this operation gives you information about the pixels which will be isolated for the chosen thresholds (some are not part of the defect).

Using thresholds D and E, use the slicing method from section 5.2. You should succeed in isolating the defect, and probably some noise around it and elsewhere in the picture.

You can then get an estimate of the defect area by summing all pixels in buffer 1 and dividing the result by 255 (using the small calculator). A problem exists: this area is probably overestimated, because pixels that do not belong to the defect have been isolated and counted in the estimation.

Now, if you have time, you can repeat all operations with a fillet having the skin residue. It may be necessary to change the lens aperture; if it is needed, take note of the previous lens aperture in order to compare it with the new one: it will be possible to see the differences between sensor requirements for the detection of two different fillet defects.

After this experiment, it is obvious that it is easier to use a white (or pale colored) background to isolate a defect from a fish fillet. But, even with a white background, the result is not perfect.

To improve these methods for finding defects, two interesting directions are worth exploring:

Enhance the contrast between the fillet, the defect and the background, through the use of filters; the aim is to obtain more distinguishable peaks in the histogram.

Reduce the area in which we are looking for a defect inside the fillet.

7. INTRODUCTION OF FILTERS IN THE ILLUMINATION AND SENSING SYSTEM

Improving the histogram quality through the introduction of filters or through an adapted background choice makes it possible to avoid time-consuming software operations (like window operations) during the isolation of a defect. Therefore the choice of the colors of the background (conveyor) and of the filters is very important.

Using the passthru mode to display the result on the video monitor, try to obtain an image with a good contrast between the normal flesh, the blood spot, and the background, by varying the different parameters:

Background color
Illumination system (lamp and diffuser)
Color of the filter in front of the light source (if the desk lamp is used)
Camera filter color and camera lens aperture

The most interesting combination is:

White background
Umbrellalike reflector and gas discharge HQL–R DE LUXE OSRAM lamp
No filter in front of the light source
Green filter before the camera lens

In fact, the green wavelength is the one whose reflection differentiates efficiently the normal flesh, the blood spots, and the white background.

8. WINDOW TO DETECT DEFECTS

If you have a case where you are obliged to use a rather dark background (like the reddish-brown one), it is really difficult to isolate the defect: There is always a certain number of pixels of the background that have the same value as the defect pixels. It is then possible to use windows in order to scan only flesh and isolate exactly the defect.

To place the windows over fish only, specific algorithms can be used, which will "fill out" the fillet contour with rectangular areas, eventually overlapping. You can then work in a specified region (window), which only includes fish flesh. Therefore, you can manage with filters and lens aperture to obtain a good contrast between the fillet and a possible defect, even if the remainder of the image (background) is saturated. In this way, you should obtain an histogram inside the window with two separated peaks for normal flesh and an eventual defect.

Figure 5 shows the scheme of rectangle windows which could be explored. Using this kind of window allows you to estimate with a certain precision the area of a defect, as a key-number for decision making, without counting in it excess pixels from the background which would have the same value as defect pixels.

Now:

Use the reddish-brown background.

Acquire images of a defect fillet with the RECORD.TP macro; make sure that the conditions described in section 5.1 are verified except the "unsaturated condition" (the background can be saturated, but not the fillet).

Execute the FINDWIND.TP macro (find the window); move the cursor inside the fish fillet in the video monitor image in order to determine the location for a window inside the fillet including the defect (you need to know the upper left angle, the number of rows and columns of the window).

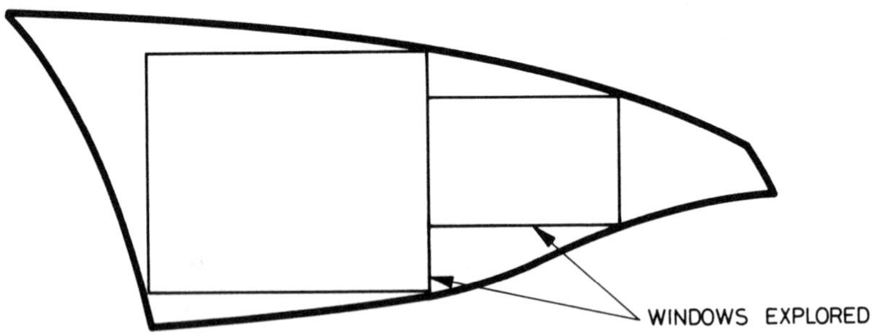

Figure 5 Scheme of rectangle windows.

Create the window (with flesh only in it).
Histogram the window (in one of both buffers); you should obtain an histogram with two peaks (one for normal flesh, one for the defect).
Threshold the buffer 0 using a threshold chosen between the two peaks:
Offset the buffer by the negative value of the threshold
Execute the THRES.TP macro to achieve the thresholding operation
Histogram the window in buffer 0 to verify that all values in it are 0 or 255.
Sum all pixel values in the window; dividing the number you obtain by 255 gives you the exact number of pixels of the defect, i.e., the defect area, key number for the further processing of the fillet.

Note: This method will not give any success in the detection of defects on the edges of the fish fillets; this is an important restriction. Therefore, it is interesting to look for other methods, or to find a way to use a white background, which makes the time-consuming windowing operations unnecessary.

9. EXTRACTION OF EDGES

9.1 Using Highpass and Lowpass Filters

Use the RECORD.TP macro to register an image as described at the beginning of the paragraph 5.1. The HIGHLOW.TP (*high*pass, *low*pass filters) macro shows this first method for isolating edges; notice that two buffers are necessary in this method.

You can repeat with an image recorded as described at the beginning of sections 5.1 to obtain the edges of a defect.

Note: The convolution operations are very time consuming when no special purpose electronic board is available. Here, there is a special coprocessor speeding up this operation. The processing time is reduced from about 1 hour to 3 minutes. Some systems are equipped with a pipeline processor that permits one to perform real-time convolution operations within 1/30 second.

9.2 Using Laplacian Filter

Use the RECORD.TP macro to acquire the same image as previously.

Binarize it in order to isolate the fillet.

Execute the BINLAPLA.TP macro (*bin*arized image, *lapla*cian filter) to apply this second method for edge extraction.

When you have finished, leave IRIS Tutor by typing:

EXIT

10. APPENDIX: MACRO LISTINGS

```
! MACRO TRAINING
! ==============

! This macro shows some very current and easy manipulations.

DISPLAY CAMERA
! You wanted to display the video input but...
! the frame grabber has to receive an external synchronization
! signal from the input source in order to receive the frames.
PAUSE

SYNC EXTERNAL
DISPLAY CAMERA
! Better now, no?
! By default, the input and output LUT are the LUTs number 0.
PAUSE

ACQUIRE 1 FRAME TO BUFFER 0
! The acquisition of this image terminated the passthru mode,
! but the acquired image is displayed on the video monitor.
PAUSE

DISPLAY CAMERA
! Vary lens aperture of the camera and see effects...
PAUSE
! Vary focusing of the camera and see consequences...
PAUSE

SELECT INPUT TABLE 6
! You are now using a thresholding input LUT ( threshold 128 )
! in order to obtain binarized images.
! Vary lens aperture of the camera and see effects...
PAUSE
! Vary focusing of the camera and see consequences...
PAUSE

ACQ 1 TO 1
PAUSE
DISPLAY BUFFER 1
! displays permanently buffer 1 using the specified LUTs.
! The following command of the macro is "HISTOGRAM BUFFER 1"; you will only
! see its effects
PAUSE
HISTOGRAM BUFFER 1
! see the number of occurences of each pixel value of the image
! stored in buffer 1. The 0 value is on the left and each vertical
! bar represents 4 pixel values (each d represents 16 pixel values)
PAUSE

ATTRIBUTE RED 250 GREEN 0 BLUE 0
! selects red as the default attribute in order to modify LUT 0
PAUSE
ALTER OUTPUT TABLE 0 ENTRY 255
! The output LUT 0 displays the pixels with value 255 in red
PAUSE
HISTOGRAM 1
! but there is no change in the histogram
PAUSE

OFFSET BUFFER 1 BY 100
```

```
! adds 100 to each pixel value
PAUSE
SET WINDOW ROW 100 COLUMN 150 SIZE_OF 200 ROWS 250 COLUMNS
! defines a certain area (AOI: Area Of Interest) as a window.
PAUSE
MULTIPLY WINDOW_IN BUFFER 1 BY 2
! multiplies the pixel values into the window by 2
PAUSE
HISTOGRAM 1
! There are now three peaks in the histogram.
PAUSE

ATTRIBUTE 0 0 255
! selects blue colour as default attribute
ALTER OUTPUT TABLE 0 ENTRY 150 THRU 250
! output table 0 displays in blue all pixels with value between
! 150 and 250, i.e. the medium peak of the histogram
```

```
! macro INITIAL
! ============

! initializes board and displays the input source.

RESET
SYNC EXTERNAL
DISPLAY CAMERA
```

```
! macro RECORD
! ============

! acquires an image, stored in both buffers 1 and 0, displays buffer 0,
! and performs an histogram operation

DISPLAY CAMERA
! please verify on the video monitor which image you are going to record;
PAUSE
ACQ 1 TO 0
COPY BUFFER 0 TO BUFFER 1
DISPLAY 0
PAUSE
! histogram buffer 0.
HISTOGRAM 0
```

```
! macro THRES
! ===========

! After offset, ends the thresholding of the image in buffer 0;
!  then displays the binarized images, and performs an XOR
! operation with a value of 1, to invert colours in the binarized image.

MULTIPLY BUFFER 0 BY 255
DISPLAY 0
PAUSE
XOR 0 WITH VALUE_OF 255
```

```
! macro SLICING
! =============

! After offset, ends the thresholding of both images in buffers 0
! and 1; then displays the binarized images, and performs an XOR
! operation, whose result is stored in buffer 1, hereby displayed.

MULTIPLY BUFFER 0 BY 255
MULTIPLY 1 BY 255
DISPLAY 0
PAUSE
DISPLAY 1
PAUSE
XOR 0 WITH 1
```

```
! macro FINDWIND
! ==============

! is a macro to help you and find the limits of a window you want
! to create. To create a window, you need to locate its upper left
! hand corner, and to know the number of lines and columns in it.

CURSOR ON
! please move cursor to find and take note of the limits of the
! window you want to create.
PLACE_CURSOR
```

```
! macro HIGHLOW
! =============

! This is one method to find the edges in an image stored in both
! buffers 1 and 0.

! Enhancement of high pixel values; BE PATIENT, IT LASTS ABOUT THREE MINUTES.
CONVOLVE BUFFER 0 WITH HIGHPASS
PAUSE
! Inverse operation; BE PATIENT, IT TAKES ABOUT THREE MINUTES.
CONVOLVE BUFFER 1 WITH LOWPASS
PAUSE
SUBTRACT BUFFER 1 FROM BUFFER 0
PAUSE
MULTIPLY 0 BY 10
! To enhance contrast.
```

```
! macro BINLAPLA
! ===============

! this is one method to determine the edges in a binarized image in
! buffer 0.

DISPLAY BUFFER 0
PAUSE
! To enhance the edges...BE PATIENT, IT TAKES ABOUT THREE MINUTES!
CONVOLVE 0 WITH LAPLACIAN
PAUSE
CONVOLVE 0 WITH LOWPASS
! To eliminate some noise in the edges.
```

REFERENCE

1. *User Guidelines for DT 2851 Frame Grabber and IRIS Tutor*, Data Translation Inc., Mass., 1988.

12
Fish and Fish Product Sorting

HÖRDUR ARNARSON *Marel HF, Reykjavik, Iceland*

1. Introduction
2. Computer Vision Sorting Versus Other Alternatives
3. Classes of Applications
   3.1 Fish Product Sorting
   3.2 Fish Sorting
4. Case Study: Length Estimation of Cod Fish
   4.1 Algorithms
   4.2 Results
   4.3 Factors Influencing the Accuracy
5. Summary
   References

1. INTRODUCTION

Whereas the use of computer vision of the detection of internal defects in fish has interested both researchers and the fishing industry for some time (1), the possibilities of fish and fish and fish product sorting have not yet received the same attention. The interest for inspection is caused by the hope for a fast payback, because of the labor-intensive manual inspection performed in the fishing industry. But problems in designing the sensing part of the inspection system (1) have resulted in interest turning more to realistic applications like fish and fish product sorting and other kinds of handling of fish products.

Although sorting is claimed to be more realistic than inspection, it is by no means simple. The fast speed, many fish species to be sorted, the batch flow in the production lines, the harsh environment in the factories, variations in form and size of each species, variations in optical characteristics of the fish as well as the elastic nature of the fish result in serious problems applying computer vision in fish and fish product sorting.

At the processing level, the machines (filleting and head cutters) still cannot handle in one batch diverse fish types or sizes without reduction in yield. Any setup due to such variations is both time consuming and costly.

At the consumer level there is also often a demand for products fulfilling both weight and form restrictions.

The level of complexity for a vision system for the fish and fish product sorting is determined by the feeding part of the system. Figure 1 lists the different levels of complexity in a vision-based sorting system.

In the fishing industry, practical levels for a sorting system are levels 2–4. Level 1 is excluded because of the orientation demand, which is unrealistic for an elastic and live object like a fish.

Level	Feeding system	Multi processing
1	Objects oriented and there is a minimum distance between objects.	No
2	Objects not oriented and there is a minimum distance between objects.	No
3	Objects oriented, no minimum distance, not overlapping.	No
4	Objects random and not overlapping.	Yes
5	Objects random and overlapping.	Yes

Figure 1 Levels of complexity in the vision part of a sorting system.

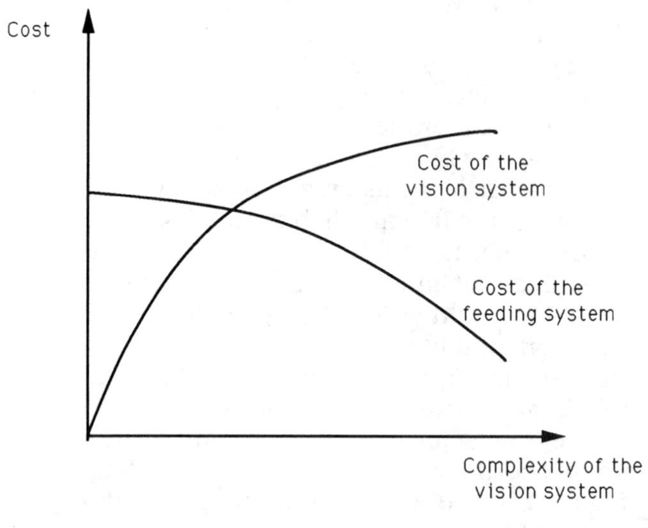

Figure 2 Complexity of the vision system vs. cost of the vision and feeding systems.

Level 5 is excluded because it normally results in high sorting errors.

The compromise that has to made when selecting a working level for a sorting system lies between the requirements (cost) on the feeding system and the complexity and speed of the algorithm (Fig. 2).

2. COMPUTER VISION SORTING VERSUS OTHER ALTERNATIVES

Nowadays fish and fish product sorting is often done manually. The seaman sorts the fish by species and size, often in four or five size groups/species. While sorting by species is done with high efficiency, size sorting is inaccurate, dependent on staff, catch, time of day, and several other factors. With this kind of sorting, little information is obtained to be used later in the factory for production control. In the factory manual sorting is again often used for both the fish and fish products.

The equipment used for sorting raw fish in the factory is mainly based on scales, although some advanced factories use length sorting equipment. The scales used are usually static scales, but the use of dynamic scales is starting. The speed of these graders is 50–75 objects/minute for the static grader and 60–120 objects/minute for the dynamic graders. The accuracy of the equipment is at best around 0.1% (2). The feeding of these graders is done manually, at level 2 in Figure 1. The fish is presented one at a time, not oriented.

Length estimation systems available on the market are based on Light Emitting Diodes (LED)(3). These systems work in level 1 in Figure 1, which is not practical because of the elastic nature of the fish. The length measurement using LEDs can only be used for a rough length estimation (3).

Among the advantages is using computer vision in a sorting system in the fishing industry are:

1. The use of computer vision sorting is labor saving because it is possible to sort by factors that otherwise only can be measured by humans. The labor saving gained

by the use of vision systems often leads to further rationalizing in the factory.

2. With the use of computer vision it is possible to lower the demand on the feeding system and work on level 4 of Figure 1 instead of levels 1 or 2 as the commercial available equipment does.
3. The speed in computer vision sorting systems is higher than in other systems (120–130 objects/minute depending on the complexity of the sorting).
4. Accurate length estimation is achieved.

Some of the drawbacks are:

1. Cost: Compared to sorting by scales or LEDs, computer vision equipment is expensive.
2. Complexity: The vision systems are more complex and less robust than other sorting system.
3. Accuracy: Weight estimation accuracy is less by a vision system than in a scale-based system because of the nature of the fish.

When discussing the use of vision systems in the fishing industry, it is important to compare it to the methods used currently and not try to solve all possible problems perfectly. It is often a trend to overestimate the performance of the existing equipment, which makes comparison difficult.

3. CLASSES OF APPLICATIONS

3.1 Fish Product Sorting

The fish product machines produce even within a given batch highly irregular products (typically fillets, but also waterjet cut fish cubes or slices). This is essentially due to the high speed mechanical handling of nonsolid meat, resulting in irregular shapes, lengths, thickness, as well as in fillets folded together, overlapping, etc.

The sorting operation consists of dispatching the fish parts according to shape and size compliance.

Required Performance The sorting speed is dictated up front by the output rate from filleting, waterjet cutting, or other machines. Such speeds are thus per line at most 250 objects/minute. The size measurement accuracy must be 2% for area and 5% for thickness.

Illumination For the estimation of length and shape, either diffused front lighting or backlighting can be used. Front lighting requires a careful selected background to get discrimination between the fish and the background. It appears that backlighting is best for all species. On the other hand, backlighting needs an evenly translucent conveyor belt, which is difficult to obtain in practice; also, the translucent nature of the fish meat causes light to pass through it.

If thickness is needed additional lateral sensor or structured lighting must be used.

Algorithm Simple image segmentation procedures, which draw minimal nonoverlapping boxes around each fish product and characterize the box in 2-D or 3-D, are sufficient. The box limits can be estimated by very high speed line scanning perpendicular to the transport direction. The acceptance rules which classify the products use knowledge about sorting criteria and about the upfront machine: This opens the way for the use of a knowledge-based system to set and update quality acceptance criteria.

3.2 Fish Sorting

Living Fish in Water

Sorting fish in water is required in fish farms (in ponds or in sea enclosures) to count fish, estimate their length and/or weight, and to estimate the weight growth curve of fish. In this application the individual fish cannot be tagged as sampling takes normally place in terms of the imaging process.

Needed Performance For sorting fish in water, the required speed is approximately 2000 fish/hour/sensor (9). This involves fish separation by gates and water flow control in the inspection tunnel.

Illumination For sorting living fish in water, contrast backlighting through water is the best method, but two points should be considered in order to find a compromise between

Fish dynamics—fish try to stay in the place where the light is if the water flow is not strong enough.

Turbulence of water—if the water flow is not laminar it interferes with the light.

Algorithms Consists mainly in selecting scan lines and extracting two-dimension characteristics.

Fish Sorting in Air

The aim of this operation is to group fish by species, length, or weight for further processing.

Sorting dead or dying fish is required on board fishing boats for packaging and storage purposes. Typically, the catch must be sorted by species, length, or weight before they go into boxes, the contents of which are compatible with an auctioning process. If fish processing or freezing takes place on board, the sorting requirements are still maintained.

Sorting by Species Nowadays this sorting is only performed manually. It might seem easy to use computer vision to sort between species (4) like cod, flounder, and ocean perch (red fish), but in the reality the species caught at the same time are often similar in shape because of the catching place, process, and net geometry. Similarities between species and big variations within each species make it difficult to solve this in a practical way using computer vision.

Integration into the existing production line also limits the usefulness of this application.

Sorting by Size There are several reasons for sorting whole fish by size (5):

1. To find the weight of a lot to get the price.
2. To direct fish within certain length/weight intervals to the proper machinery, e.g., head cutters and filleting machines.
3. To find the weight before and after head cutting and filleting for yield control.
4. To select the fish for the production.
5. To be able to predict the amount of meat on the fish for reasons 3 and 4.

The most likely parameters to give an estimate on the amount of meat are: Weight, volume, area, length, and thickness.

1. Weight can be measured accurately (0.1%) with digital scales (2).
2. Volume gives the same information as the weight because of the equation:

$$\text{weight} = k \times \text{volume}$$

   k has been shown to be constant (5) independent of the amount of meat on the fish. When the fish has been gutted, the belly and the belly flaps make it difficult to measure the volume accurately. The round shape of the fish cause errors because the methods which are used only measure in 2.5-D.
3. Area: When working with gutted fish, it is difficult to measure the area accurately because of the belly flaps and fins. Figure 3 shows results from area estimation of gutted fish with the belly flaps and fins in different positions. A 30% variation (compared to the biggest area) in the 2-D projected area measured by a camera was observed, 16% because of the belly and 14% because of the fins and tail.
4. Length can be determined accurately by using computer vision. The effect of different positions of the belly flaps and fins disturbs the length estimation, but this can be reduced using an effective algorithm (section 4, case study).
5. Thickness: It is difficult to define the thickness of the fish in a way such that a vision system can measure it on a moving conveyor.

Required Performance The supply rates are highly irregular, but an average requirement is 100 fish/minute.

The needed accuracy is highly dependant on each application, but is approximately ± 1 cm in length estimation, ± 10% in area and volume estimations, ± 0.2 cm in thickness.

Illumination For sorting fish in air, either diffused direct light or backlighting could be used. Usually backlighting causes fewer problems except for thin fish, because of light going through the fish. Direct light requires a special setup to reduce glinting and to

	Fish 1		Fish 2	
Measure-ment (No)	Area (pixels)	Length (cm)	Area (pixels)	Length (cm)
1	26447	60.6	24791	57.1
2	22144	61.4	21385	56.7
3	18462	59.8	18730	57.4
Variation				
1	100%	100%	100%	100%
2	86%	101%	86%	99%
3	70%	99%	76%	101%

Figure 3 Variations in length and area due to different positions of belly flaps and fins.

avoid shadows even on specimens with no scales because of the silvery belly of the fish.

If thickness is needed, an additional lateral sensor or structured lighting must be used.

Algorithms Classical high speed sorting procedures (6,7) apply for moving parts. They involve picture or linescan binarization, and exploit the shape of the fish to generate a fish characterization by a variable number of 2-D or 3-D slices at known intervals in space and time. These slices are obtained by sensor scan control in connection with the transport irrespective of fish orientation. Sorting by species as well as precise length estimation involves hypothetical head and tail location each time an object is detected and carrying out morphological filtering of selected windows for feature detection (head, tail, fins).

4. CASE STUDY: LENGTH ESTIMATION OF COD FISH

In cooperation between the AI-Vision group at the Technical University of Denmark (TUD), the Research Laboratory of the Danish Fishing Ministry (FF) in Denmark, and Marel hf. in Iceland, a prototype for length estimation of cod fish was built (Fig. 4) and tested in the fishing industry in Denmark. The prototype consists of a 3-m-long conveyor with a built-in light source (diffused backlighting) and a wooden box to house the camera and shield the part of the conveyor where the fish was imaged.

The system works in level 4 in Figure 1 as the feeding system provides objects randomly but not overlapping.

The length was defined as the line going from the middle of the tail to the top of the head on a straight fish (Fig. 5). The algorithm allows the fish to be bent (Fig. 6).

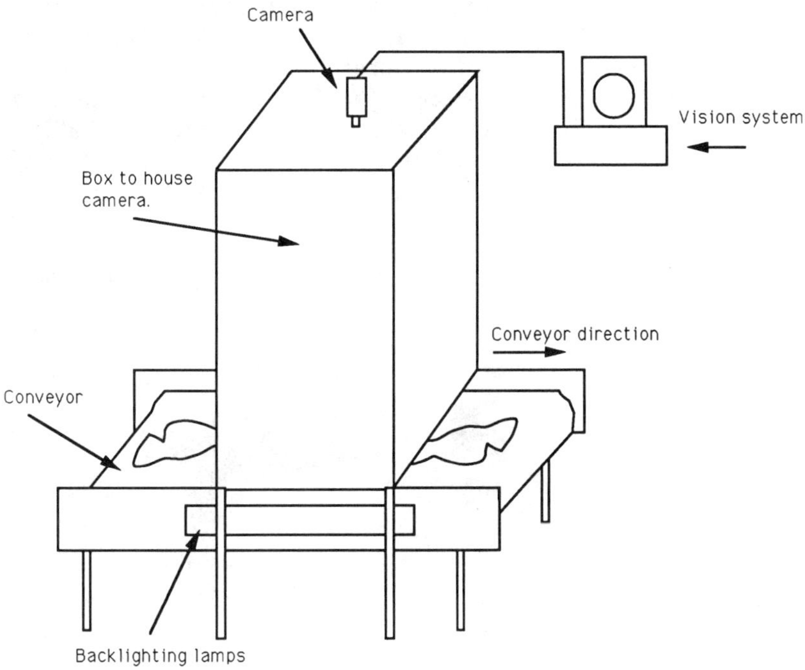

Figure 4. The test setup.

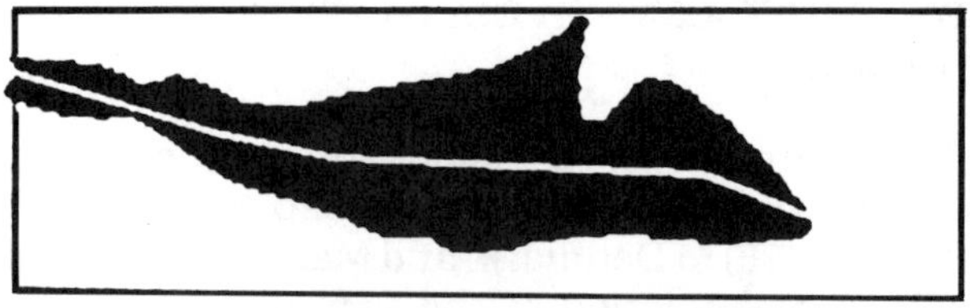

Figure 5 Straight fish; length estimation line.

The accuracy of the computer vision–based length estimation was found to be ±0.9 cm (two standard deviations) when correlated against manual measurement; this accuracy was size independent and orientation independent when the fish was lying within a 75° cone from the transport direction.

4.1 Algorithms

The programs (Fig. 7) were written in FORTRAN on a MicroVax-II VMS host with a ITI200 vision system from Imaging Technology Inc. using the ITEX subroutine library.

A built-in setup procedure adjusts for variations in lighting and background. Two fish can enter the filed of view (FOV) at the same time and bent fish are allowed. Blurr is reduced to one pixel (Fig. 8) independent of conveyor speed using morphological filtering (8).

The algorithm detects fish arrivals estimates a 2-D box encircling the fish, scans the fish, finds head, tail, and belly, meas-

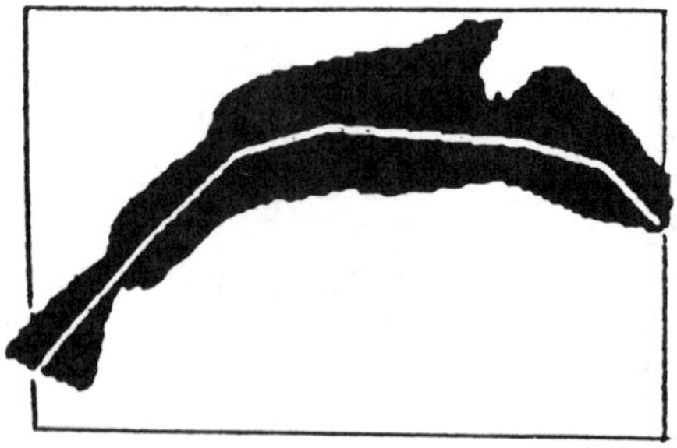

Figure 6. Bent fish; length estimation line.

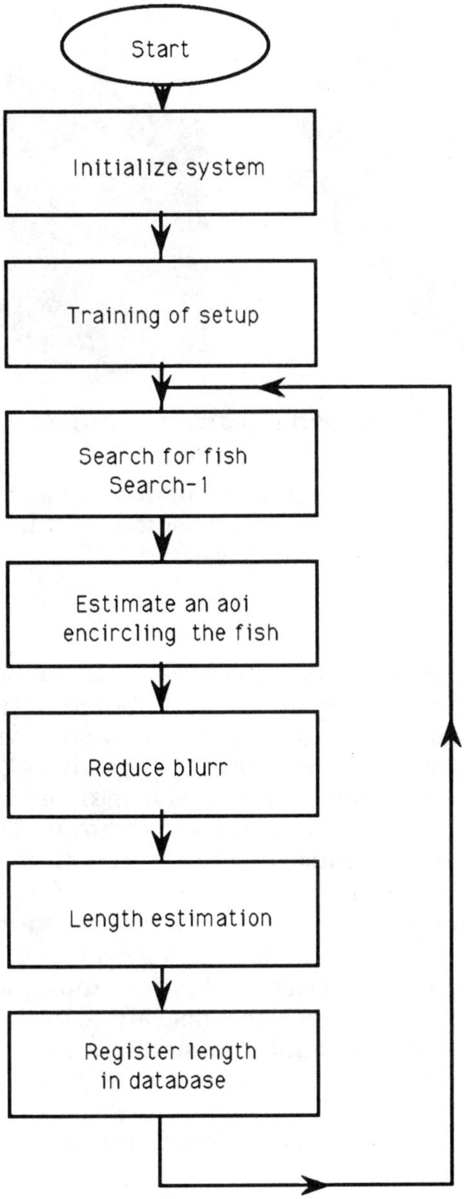

Figure 7 Length estimation algorithm.

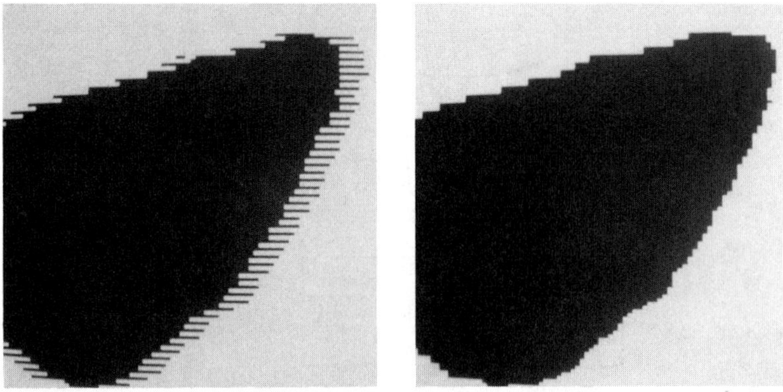

Figure 8 Reducing blurr using morphomathematical filtering.

ures the length of the fish from the middle of the tail to the top of the head, taking into account the position of the belly, and finally registers fish length in a database.

4.2 Results

The setup was tested during one week in September 1988 in the Bagenkop fishing factory in Denmark. Seven hundred and fifty fish were measured manually and by computer vision. Results from this tests are found in Figure 9, where a histogram of the difference between manually measured length and a computer vision–measured length of each cod is shown. This difference shown in Figure 9 is caused by errors in both the manual and computer vision measurements.

To test the accuracy of the manual measurement, 50 fish were measured twice. The difference between these measurements had a standard deviation of 0.5 cm and a mean of 0.4 cm. Taking this error and other errors not caused by the computer vision equipment into account, the variations in the computer vision measurement were estimated to have a standard deviation of 0.45 cm.

The calculations speed for length estimation was between 0.55 and 0.8 s/cod fish.

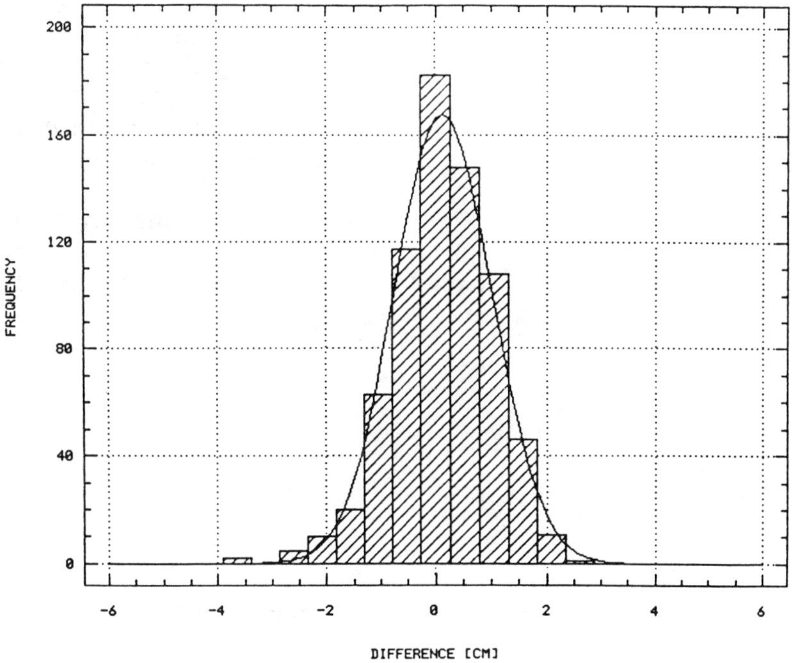

Figure 9 Difference between manual and computer vision–measured length.

4.3 Factors Influencing the Accuracy

The factors influencing the accuracy of the computer vision length estimation are shown in Figure 10. The typical and minimum errors caused by each factor assuming a field view or 1.0 × 0.66 m. From Figure 10 it can be seen that it is difficult to get better than 0.9 cm accuracy on the length estimation by computer vision, but this accuracy is also possible to obtain in manual length measurements on gutted cod fish.

5. SUMMARY

The use of computer vision in fish and fish product sorting shows great promise. There will always be need to compromise between the complexity of the vision system and the demands on the feeding system. Integration into existing production lines is also a problem.

In this chapter a flexible system for length estimation of cod fish is presented. This vision-based system works with an accuracy of ±0.9 cm and with a speed of 90 fish/minute. Fish are allowed side by side and bent.

Error type	Typical at object (mm)	Minimum at object (mm)
Position of length estimation line	50	0
Camera resolution	0.9	0.5
Capacity of frame buffer	0.9	0
Blurr	20	1.8
Optics	2.5	1

Figure 10 Factors influencing the accuracy of computer vision–based length estimation of cod fish.

ACKNOWLEDGMENTS

The author would like to thank Dr. L. F. Pau, AI-Vision group at Technical University of Denmark, Dr. J. Heldbo at the Research Laboratory of the Danish Fishing Ministry, and the people at the Bagenkop fishing factory for their cooperation in this project.

REFERENCES

1. Arnarson, H., K. Bengoetxea, and L.F. Pau. 1988. Vision applications in the fishing and the fish product industries, *International Journal of Pattern Recognition and Artificial Intelligence*, 2(4), 657–669.
2. Marel hf, Höfðabakka 9, I5-111, Reykjavik.
3. Com Data A/S, P.O. Box 19, FR 700, Klakksvik.
4. Muus, B.J., and P. Dahlstrøm. 1974. Collins Guide to the Sea Fishes of Britain and North Western Europe, *Collins, London.*
5. Heldbo, J. Erhvervsforsker report, Fiskerriministeriets Forsøgslaboratorium, Copenhagen, to be published.
6. Bretschi, J. 1981. Automated Inspections Systems for Industry, IFS, Oxford, UK.
7. Pau, L.F. 1984. Approaches to Industrial Image Processing and Their Limitations, *IEE Transactions on Electronic & Power*, February 1984, pp. 135–140.
8. Giardina, C.R., and E. R. Dougherty. 1988. Morphological Methods in Image and Signal Processing, Prentice Hall, Englewood Cliffs, NJ.
9. Pau, L.F. 1988. Vision for fish inspection and sorting, *Proc. Conference on Data Processing for the Fishing Industry*, Nordic Industry Fond, Oslo, Norway.

13

Preliminary Results of a Computer Imaging System Applied to Estimating the Quantity of Larvae and Fingerling Fish for Aquaculture

MICHAEL D. SPRATT *Applying Evolving Technologies, Montpellier, France*

1. Imaging Systems Applied to Aquaculture
2. Materials and Methods
   2.1 Larvae, Fish, and Shrimp
   2.2 Containers
   2.3 Ligthing
   2.4 Objectives
   2.5 Camera
   2.6 Imaging System
   2.7 Computer
   2.8 Programs
3. Conclusion
4. Discussion
   References

1. IMAGING SYSTEMS APPLIED TO AQUACULTURE

In fish hatcheries large quantities (5,000–20,000) of larvae and fish are counted on a regular basis. The counting is done during certain stages of development, most often when they are being transferred from one basin to another or just before the fish leave the hatchery. At present counting is performed by photoelectric chambers or manually (one fish at a time). We have performed some experiments using an imaging system, personal computer, and programs designed to "estimate" the number of larvae and fingerlings in small basins.

Counting of eggs, larvae, and fish via computer is not a new or an untried problem (1–3). Mabo Aqualine A.S. of Norway manufactures the Ma-Bo Aqualine Fish Counter, which uses an optoelectronic photo register to count the number of fish passing through the mechanism.

In most cases manual counting is still the preferred method. Our statistics indicate that one person can count between 10,000 and 12,000 larvae a day. An average-size hatchery must count between 40,000 and 60,000 larvae four to six times a year. Thus, each year a medium-sized hatchery spends approximately 8,000–10,000 FF to count fish (4 persons × 50FF/hour × 8 hours × 5 times per year). Manual counting is also a very tedious, tiring, and time-consuming task. This method also stresses the fish, which often causes disease and death of the young larvae. Larger hatcheries only estimate the number of fish.

Often an experienced person tries to estimate the number of fish by considering the volume (i.e., size of the basin, surface area, water depth) and the size and density of fish. A person might also consider the density of fish in the middle and around the edges. This method, as reported by the users, is estimated to have a 3–5% margin of error. In the absence of contradictory data, we will assume that this margin of error is correct.

This method seems to us the simplest, easiest, fastest, and least expensive approach to estimate the number of larvae present in a basin. We therefore tried to keep our automated method as simple as our advanced human model and its method of estimating fish. Thus, our objective was to design a system that estimated the number of fish with about a 5% margin of error and that did not require a lot of sophisticated and expensive equipment and programming. In addition, we wanted a system that was programmably universal for aquaculture installations, easy to use, dependable, and, if possible, affordable.

Our system is simple. An image containing larvae or small fish is acquired. A lookup table (LUT) divides the image into three grey levels: black (00), grey (128), and white (255). The program counts the number of pixels having the values 00, 128, 255, it then searches a database containing these values and the number of larvae associated with them the closest correlation provides an estimation of the number of larvae represent in the image. The imaging problem and application is a basic one common in many industrial applications in which objects are counted and their size measured. It is a system where texture and statistics are the basic program modules.

There is, however, much more to estimating fish than just counting black pixels and dividing by the number of pixels required to make a fish or searching in a database for similarities. There are several factors, considerations, and information which make this method an interesting approach to inventorying fish.

For example the percentage of black can indicate the number of fish also. As fish are added, the percentage of black augments at the expenses of the white. An adverse effect of this is that fish tend to superimpose as their density is increased, thus reducing the number of black pixels, their percentage, and the average number of pixels per fish. Thus, an "optimal" image should contain percentages of black, grey, and white and ratios black to white (black/(black + white)), black to grey (black/(black + grey)), grey to white (grey/(grey + white)), and image noise (white/(white + grey)) within specified limits. These values will also indicate the quality of the image and the number of fish present, which in conjunction with established limits (allowable minimum and maximum number of fish in a basin) should increase the precision of this method notably.

Using a variety of image processing and utility programs we established the general limits for many of these variables and their relationship, but the precise limits, according to the hardware configuration, program variables and biological elements vary and currently remain unknown.

2. MATERIALS AND METHODS

The material used was as shown in Figure 1. Although we tried several different hardware configurations, the basic system remains the same:

1. A basin large enough to hold a quantity of fish without overcrowding and stressing, containing only enough water to allow free motion but which reduces the possibility of the fish superimposing themselves.
2. A camera positioned in such a way that the entire basin is visible and connected to a computer imaging system.
3. Fish are placed in the basin, and an image is acquired.
4. The processing program binarizes the image and counts the number of pixels, performs the calculations, and estimates the total number of fish.
5. The basin is emptied, the next lot of fish are placed in the basin, and the above steps are repeated.

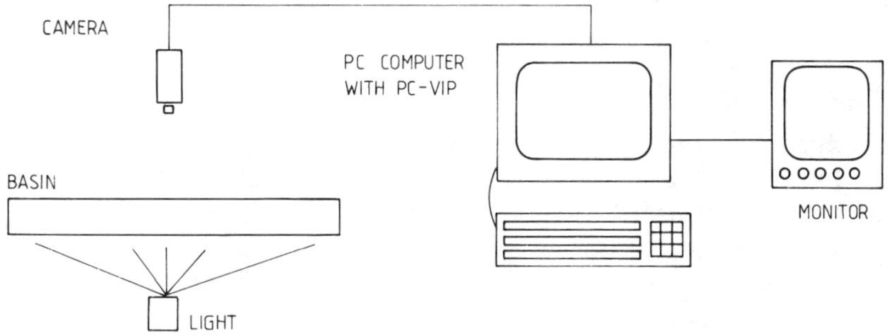

Figure 1 System configuration.

This system is not limited to counting larvae and fish but can include shrimp, mussels, and microscopic organisms. Thus, many variables, i.e., subjects, the basin size, and camera location, would take into account these considerations and a flexible device should be constructed.

2.1 Larvae, Fish, and Shrimp

The fish used in our experiments were ornamental aquarium fish, *Zebra danio* (3 cm) an *Tilapia buttikofeu* (1 cm and 3 cm). Several experiments with shrimp (*Peneus japonicus*) were attempted, but the results were inconclusive.

Fish were added one at a time or in groups of 5 or 10 and the system allowed to acquire images and store the results. As many as 75 (all that were currently available) 1 cm *T. buttikofeu* were placed into a basin 11.5 cm long × 8 cm wide × 5.5 cm deep. The water level was kept low (between 2 and 4 cm) in order to minimize reflections, light dissipation, and superimposing of fish upon one another.

An interesting observation concerning the *Z. danio* and *T. buttikofeu* is their use of space. Both species have a tendency to group, thus exploiting only a fraction of available space. The density, as with most captive fish, is controlled by the size of the basin. The optimal or preferred density is that of the fish in an unlimited area such as a lake or ocean (5–9). In this natural environment fish establish optimal distances from their conspecifics for reasons such as social interaction, mutual protection, etc.

Using an imaging system one might find the density preferred by fish and use this information to create an artificial environment that provides the dimensions "optimal" for a given number of the same species.

Experiments with Shrimp (Peneus japonicus)

Having access to an abundance of 2–4 cm juvenile *P. japonicus*, we took our chances and tried to count them also. After several tries it was found that the best contrast was obtained using a transparent blue or green background. This caused the thorax and eyes to stand out as black after the LUT had been programmed. An interesting observation is that shrimp have a tendency to remain apart from each other, possibly because of cannibalistic tendencies. This behavior could indicate that our approach toward counting larvae might be applicable to shrimp.

2.2 Containers

The containers we used were a variety of small, white translucent Tupperware and plastic containers. Several small counting cubits, aquariums, and miscellaneous containers were also tried. We propose containers of different sizes for different types of fish at different stages of development.

The following is a list of containers used which had potential:

No.	Description	Material	Length (cm)	Width (cm)	Depth (cm)	Color
1	Small Tupperware	Plastic	11.5	8	5.5	white
2	Large Tupperware	Plastic	14	10	6.5	white
3	Medium tray	Plastic	37	21.5	7	white
4	Large tray	Plastic	47	32	8	white
5	Aquarium	Glass	50	50	50	clear
6	Small cubit	Plastic	13	3	12.5	clear
7	Large cubit	Plastic	20	5.5	16	clear

In all cases the containers appeared to have a homogeneous background, but using the imaging system to analyze the pixel values, it was found that slight variations in tint were common at the edges of the basins. In every plastic basin tested, the background became increasingly darker with wider variations in pixel values occurring near the edges, probably caused by the presences of the container's sides merging with the bottom. This region contains anomalies in material consistency and density, which provokes an uneven dissipation of light through the plastic which effects the pixel's value. This is further enhanced by the objective (see Sec. 2.4.), which also reflects light around its edges.

2.3 Lighting

The lighting system used with the most success was an "underlighting" system for visualizing photos and slides. A series of other lighting arrangements were tried but with little success.

The underlighting system provided a very contrasted image when used in conjunction with a window and an opaque filter created by cutting a window the size of the basin in the center of a piece of plywood and placing the plywood over the light source. This effectively blocked out all light except that escaping through the window in which the basin was placed. This filter also provided a clearer and more homogeneous background with well-defined edges around the basin. The black edge also reduced the shadowing effect caused by the larvae as they approached the wall of the basin, and it provided an outline for establishing the dimensions of the window's edge.

2.4 Objectives

The use of various objectives illustrated the effects they can have upon an image such as focal length, depth of view, and diaphragm setting. These were all calibrated in order to obtain a dependable two-dimensional reproduction of a three-dimensional image. The effects that had the greatest impact were distance to subject, image focal depth, and attenuation.

Attenuation and the defractive index are important factors associated with all objectives. Around the edge of an objective, light is being dissipated away from the receptor material. This effect in itself is not too important, however, it does affect the size by increasing the number of pixels a fish will appear to have as it moves from the clearer, more brightly lit center toward the more ombré edges. The solution to this "observed growth" in pixel numbers is to reduce the window size and center it in the image.

The following were some of the objectives tested:

No.	Name	Size (mm)
1	Olympus	28–48
2	Olympus	135
3	Olympus	50

2.5 Camera

Several cameras (and a VCR) were tested in conjunction with the various objectives and lighting systems. However only two different cameras were used to capture images: the Panasonic WV-CD 50 and the Hitachi CCTV Camera HV-165. The Panasonic WV-CD 50 camera was used for the actual experiments, whereas the Hitachi CCTV Camera was used to calibrate and test the system. This tube camera was manufactured in 1968, and although it worked nicely as an image-capturing device, it lacked the modern signals standard in today's CCD cameras.

The advantage of both cameras is that the automatic gain can be inhibited in the Panasonic, and it does not exist in the Hitachi. Automatic gain is feature of most cameras which causes the camera to automatically lighten a somber image. Thus, as more larvae are added to the basin, the image becomes darker, but a camera with automatic gain will lighten the image before it transmits the signal to the imaging system.

2.6 Imaging System

The imaging system used was PC-VIP (Video Image Processor). It is a PC-based image processor designed, manufactured, and donated by VMES S.A., France.

The notable features of PC-VIP are its 1 Mbyte of video VDRAM configured in four memory plans (0,1,2,3) of 512 × 512 (262,144) pixels of 8 bits (0 to 255). The module also has 8 camera inputs, 16 lookup tables (LUT) on input, 8 LUT's on output, and pseudo-color output.

PC-VIP is sold with a variety of software programs. The principle programs used for these experiments was "LOOKER" and the "primitifs library." LOOKER is a utility program used to access PC-VIP to perform image acquisition and manipulation as well as other initial analysis and image calibration functions. The primitifs library is a combination of Microsoft C and Assembler language programs which facilitate reading and writing to memory, command, and control registers. The registers effect the choice of input cameras, the acquisition of images, the programming of LUTS, accessing of image plans, etc. Programming in C and linking to the primitifs library permits the use of these prewritten and pretested routines in user-defined applications.

2.7 Computer

Two different personal computers were used in the tests. The first system was a Tandon XT/PC computer with an Intel 8088 processor running at 4.88 MHz. It was found to be inappropriate for this application for several reasons. The second system was a Samsung AT/PC with an Intel 80286 processor running at 10 MHz. This system alleviated many of the problems encountered in the Tandon XT.

2.8 Programs

The software used in these experiments was a mixture of customized programs and "off-the-shelf" packages. The major software components were the User Interface, Image Processing, and Data Analysis. Three programming languages were used: Microsoft's Assembler and C, and Foxbase Plus (which is compatible with Dbase III). A variety of utilities and programming tools (i.e., Lotus 123 for data analysis, GENIFER—a screen and program generator) were also employed.

The user interface and database was programmed using Foxbase Plus because it permits the development of prototype systems relatively quickly. Using database management programs, such as Foxbase Plus, allows simple development of ergonomic data entry screens and menus which control, edit, and store values. These values can be passed to the C or Assembler language programs, which then manipulate the image and returns control back to the user interface. This method allows for rapid, inexpensive development of prototype applications, which are immediately usable by nonskilled users. They also provide a valuable tool, which permits the user to work while acquiring a clearer knowledge of their application, needs, wants, and likes. Modification of the menus, screens, and formulas can be effected quickly and in many cases by the users themselves, without compiling and linking the entire system. In addition, with the use of multitasking systems such as UNIX, ZENIX, and OS/2 image processing programs can be executed in background or phantom mode, thus freeing the user's screen for other applications.

The user interface had four major subprograms: control menus, data entry, data transfer, and data analysis. The control menus were used to verify the presence of required files and direct the user during data entry and updating. The data entry consisted of editing and verifying user input such as the fish's name, its age, number of times to count each basin, window size, the date, operator's name, camera, objective, basin, etc. The data transfer wrote the User Selectable Parameters (USP) to the control file from which C and Assembler programs obtained and executed their instructions.

Four USP were passed from the user interface to the image processing programs. They were camera number (0–7), number of images to acquire (0–9999), the window dimensions (0,0) × (511,511), and the threshold values for the lookup table (0–255).

As stated earlier, the PC-VIP module has eight camera inputs. The user selected the source camera by specifying the input line number of the camera. This allowed several cameras to be connected at one time and easy switching between cameras. The user also supplied the counter which determined the number of times the program looped. Thus, a user could instruct the program to acquire and process from 0 to 9999 images. In this version of the system the user had to know, by using LOOKER, the window dimensions and optimal threshold values. Future modification include allowing the system to calculate the window dimensions and establish the threshold values.

Finally the data analysis module would access the output data from the C and Assembler programs and calculate the ratio and percentages. It then searched an appropriate database for corresponding data. If no anomalies were discovered, the closest matching record in the database was selected and its corresponding number of larvae selected.

It is hoped that this method will be replaced by a heuristic motor. The storage of and access to an information base by an artificial intelligence program might aid in reducing the margin of error. By "intelligently" comparing results with past experiences, the program could "learn" to be more accurate.

The image processing program was written using Microsoft C version 5 and a combination of primitive C routines provided by VMES. The principle components were the image acquisition, the pixel counting routine, and storing the output. Thus, upon successful reception of the user-selectable parameters, the program initialized PC-VIP, programmed the LUT, started acquiring images, counted the number of black, grey, and white pixels within the window, and wrote the results to an output file.

3. CONCLUSION

The initial results of our experiments indicate that estimating larvae and fish populations is possible based upon previous assumptions and method, which requires, however, established factors related to the image (i.e., camera height, focal length, lighting, contrast, basin area, water level, etc.), factors which human observers take for granted. The biological factors such as subject size, maximum and minimum number of fish at any one time within a basin, and their behavior must also be known. And finally, statistical information like the number of black pixels per fish, percentage of black, etc., depending on the number of fish present, must also be obtained.

In order to test this proposed method, a full-scale model will be constructed (in process at this time) and tested using a variety of fish at various aquaculture installations. Most variables (camera, objective, lighting, etc.) will be fixed and remain constant. In addition results will be analyzed in greater detail to provide a clearer understanding of the fish's behavior and the optimum method for counting them.

Figure 2 shows the factors, their interactions, and their effects on the viability of this method. The graph demonstrates the dynamic interactions between the percentages of black, grey, and white, the ratios of black to white, black to grey, grey to white, and image noise, the absolute number of black pixels as the number of fish present changes. The images below the graph represent an approximation of the visual effects encountered during testing. The size of the container and the fish are relative as the objective and camera height can be configured in such a manner as to reproduce the same visual effects.

The critical factors that determine an image's integrity and thus the viability of this method are:

1. The number of images taken
2. The window area, in pixels
3. The upper left and lower right X,Y coordinates of the window

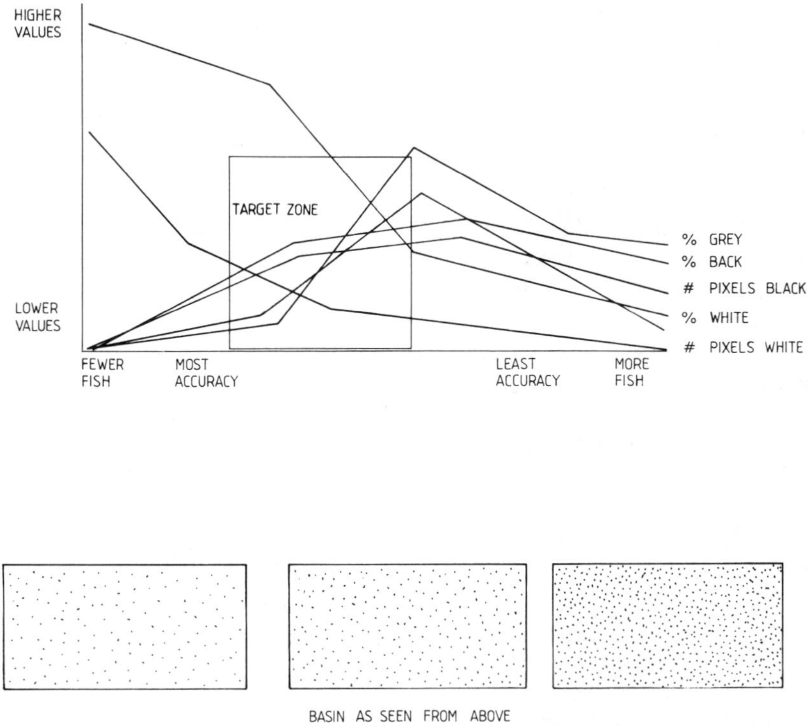

Figure 2 Observed relationship between image variables and larvae density.

4. The threshold values, lower (black), middle (grey), and upper (white)
5. The ratios black to grey, black to white, grey to white, and image noise
6. The percentages of black, grey, and white
7. Total pixels black
8. Total fish present
9. Pixels per fish

The above factors are discussed in sequence below.

1. In order to accurately calculate the number of fish in a basin, several images of the same basin should be taken and the results analyzed and averaged. This method smooths skewed data resulting from superimposed fish. The more images per basin, the more accurate the results tend to be, but the processing time is proportionally longer. We established a minimum of three images for each basin.

2. The window size has several important physical and statistical factors associated with it. The window's area affects the processing time, the amount of fish that can be placed in the basin at any one time, and the percentages and ratios. In addition, the camera height, the objective's refractive index, and diaphragm settings all impose size constraints upon the window's area.

3. The upper left and lower right (X,Y) coordinates are important as they determine the size and physical location of the window within the image.

4. The threshold values differentiate the larvae from the background. By programming a tri-level LUT, we can divide the image into fish, noise, and background. The fish in this case were black, the background white, and the grey, or noise, was neither.

5. The ratios of black to grey (black/(black + grey)), black to white (black/(black + white)), grey to white (grey/(white + grey)) and image noise (white/(white + grey)) is also calculated.

Image noise is distortion; it is regions which are neither fish nor background. They are regions in transition from fish to background or vice versa. This region cannot be considered fish or background; it is a "dead zone" which surrounds each fish but is concentrated near the fin and tail regions. The amount of the "dead zone" can be used to calculate the noise-to-image ratio, which implies a good image or not.

What is not clearly understood are the effects the type and quantity of fish have on the level of grey in the image. The grey and black regions should climb as more fish are added, while the percentage of white declines. Calmer fish or ones which maintain denser groups should have proportionally smaller regions of grey surrounding their bodies. We assume there is a point at which the percentage of each shade reaches a critical point, after which the image has little integrity.

6. The percentages of black, grey, and white imply concentrations. Each is in relation to one another and the number of subjects. Once a certain percentage of black, grey, or white has been reached, an image can no longer be considered valid.

7. The total number of black pixels (minus the number of black pixels in the control) is the simplest method to imply the number of fish percent.

8. The total number of estimated fish is the result of calculating all the above factors and selecting an appropriate estimate.

9. The number of pixels per fish is established during calibration or by using the system. Experiments indicate that the average number of pixels per fish changes in relation to the number of fish present. Thus it is necessary to test and calibrate the number of black pixels per fish throughout the possible range of fish in a basin.

4. DISCUSSION

The results from the above experiments are still basically inconclusive. There remain many questions concerning the viability of this method in nonlaboratory settings and with certain sizes and types of fish (e.g., eels, shrimp). In addition, an ongoing study of the ratios, margin of error, and their variations is also needed. Thus, it seems unlikely that a "one-size-fits-all" system is possible; however, estimating fish with this method is possible, but to what limit and with what accuracy still remain unknown.

Counting fish using the above method has several other very important factors and observations worth mentioning. First, the clients are commercial aquaculture installations and not research laboratories, and the data has commercial and management implications as well as biological value; in addition the subjects are live and have immediate economic importance. The users are nontechnical and the environment is non–computer friendly (wet, corrosive, rugged). And finally, a system of this nature must also be inexpensive. The major expenses are the initial outlay for material and training and ongoing costs of operation, maintenance, and upgrade. The capital equipment could vary considerably:

	Minumum	Maximum
Basin and lighting	1 500 FF	3 000 FF
Camera	4 500	8 000
PC imaging board	10 000	15 000
PC computer	10 000	20 000
Programs	8 000	25 000
	33 000 FF	71 000 FF

Initially a portable PC-based system, with a minimum configuration of memory, graphics, disk space, etc., might be useful during the early stages of testing and development or for use at multiple installations. The system should, however, be rugged and water and shock resistant. In addition, information such as the date, number of images per basin, number of fish per image, and the average per basin with the final total, etc., could be very useful for management and in analyzing the system's performance. Thus, a detailed report of all activities might be necessary for record keeping and system evaluation.

Once a satisfactory system is fully functional, a transfer of the programs from a PC-based system to an autonomous system with its own special configuration for counting fish seems plausible. In this system the programs could be hardwired for faster execution.

Concerning the hardware required for this application, a specialized system designed for a corrosive environment with industrial or military standards for ruggedness might be required, as well as special I/O ports for keyboard input, a small screen LED (perhaps a 1-line minimum), plus serial and printer output. In addition specialized ROM resident programs could execute much faster, thus increasing throughput.

Concerning the cameras, a standard CCD, VIDICON, or video camera without automatic gain and stable image acquisition is needed which is also rugged and resists corrosion, moisture, and shock and is, if possible, inexpensive and lightweight. An optional camera feature would be a built-in viewing screen for aligning and focusing, thus eliminating the need for a monitor.

Concerning the basin material, size and color all depends upon the type, concentration, size, and behavior of the fish as well as the lighting. The lighting should be uniform and enhance the specimen, but does not necessarily have to be in the visible spectrum.

The programs can be improved by implementing some of the following modifications:

1. Calibrate the image using a grid of known values placed in the image
2. Provide fixed points so that the program has references
3. Analyze pixel counts within separate areas of the window (each area having the same number of pixels)
4. Use a special software LUT which is area sensitive (dynamic adjustment to lighting)
5. Analyze the distribution of fish
6. Store the image and count later
7. Perform an object recognition routine
8. Perform an analysis of the axis of an object
9. Perform a surface area analysis of an object
10. "Hardwire" or "cable" the programs for faster execution
11. Fix the hardware so it cannot be changed
12. Customize the system for a specific fish, quantities, sizes, and container type

There are also several proposed methods for handling and counting the fish using this method, for example, a system whereby, during the transfer from one basin to another, the larvae must pass through a special container designed to count them. The transfer can be done manually or under the control of a computer. The container could be an integral unit, closed but with valves which open (under computer control) to allow the fish to enter. A camera mounted on the hood, which completely covers the basin, takes the image and transmits it to the computer. Once counted, the computer opens the output valve, allowing the water and fish to swim into a receiving basin, and the cycle continues until the preprogrammed number of fish have passed. Another solution is to design portable basins and a fixed lighting system in which the basins are "inserted" into a notch.

other solution is to design portable basins and a fixed lighting system in which the basins are "inserted" into a notch.

In closing, we would like to provide some observations on the management and financial aspects of developing this technology.

This method has a moderately low (initial) purchase price and next-to-nothing running costs and compares favorably against outdated expensive manual counting and handling techniques. But considering the high rate of unemployment, government subsidies for hiring the unemployed or students and minimum labor costs, manual fish counting is economical. Our experience also indicates that few fish hatcheries have both the capital and experience to invest in automated equipment. Thus, it could be difficult to justify investing in expensive computer systems which are complicated, difficult to operate, have short technical lives, and are often outdated before they are purchased.

In addition, as in any application dealing with imaging, special attention to details is required. It is important that experiments and prototype systems destined for commercial applications be thoroughly tested and calibrated to meet well-defined specifications. The system must be well documented and commercially presented. Technical support and after sales service must also be considered. The system should be kept inexpensive in order to interest commercial fish hatcheries. And finally, the professional application of developed innovative technology cannot be achieved without professional management, organization, planning, and funding.

REFERENCES

1. Tanasichuk, R. W., Armstrong, C. and Ware, D. M., 1985. An improved photo-electric fish egg counter. Can. J. Fish. Aquat. Sci., 42: 1255–1258.

2. West, D. C., Thompson, R. K. and Huron, J. M., 1983. Opto-electric counter for postlarvae and juvenile shrimp (*Machrobrachium rosenmbergii de man*). Aqua. Eng. 119–133.

3. Roff, J. and Hopcroft, R. 1986. High Precision Microcomputer Based Measuring System for Ecological Research. Can. J. Fish Aquat. Sci., 43: 2044–2048.

4. Boyar, H. C. and R. A. Clifford, 1967. An Automated Device for Counting Dry Eggs, Trans Am Fish Soc 96:361–363.

5. Serebrov, L. I., 1974. Relationship between School Density and Size of Fish. Journal of Ichthyology, 15:135–140.

6. Partridge, B., Pitcher, T., Cullen, M. and Wilson, 1980. The Three-Dimensional Structure of Fish Schools. Behavioral Ecology and Sociobiology:277–288.

7. Partridge, B., 1982. La Structure et le Rôle Social des Bancs de Poissons. Pour la Science. 92–103.

8. Pitcher, T. J., Magurran, A. E. and Edwards, U. I., 1985. Schooling Mackerel and Herring Choose Neighbors of Similar Size. Marine Biology, 86:319–322.

9. Halls, S. J., Wardle, C. S. and MacLennan, D. N., 1986. Predator Evasion in a Fish School: Test of a Model for the Fountain Effect. Marine Biology, 91:143–148.

14

Advanced Vision Methods and Technologies for the Fishing and Fish Product Industries

L. F. PAU *Technical University of Denmark, Lyngby, Denmark*

1. Introduction
2. Nonimaging Sensors for Fish and Fish Product Sorting
3. Sensor Fusion for Fish Product Inspection and Handling
4. Color Spectrum Illumination for Quality Control and Freshness Grading
5. Conclusions

References

1. INTRODUCTION

At first sight, applications of vision to the fish and fish product industry should be highly specific, owing mostly to the three-dimensional flexible nature of these objects and to difficult specular illumination problems. After further analysis, however, other items feature comparable properties (e.g., biological cells, plastic containers, gas bubbles), even if the scales involved are entirely different, and in those instances acceptable vision systems have been in use for some time; so why not extend them to fish/fish products as well?

First-generation vision applications in the fish industry exploit this apparent contradiction by technology transfer from other more mature fields; this book features several, e.g., worm detection by feature selection, blood spot detection, length estimation.

So what about the next generation systems? Properties rather typical to fish will require novel solutions or methods, which, in turn, may benefit other application fields by a reverse technology transfer. This chapter aims specifically at presenting said novel methods or technological solutions for exploitation first in the fish industry, and later elsewhere.

2. NONIMAGING SENSORS FOR FISH AND FISH PRODUCT SORTING

Second-generation fish or fish product sorting will emphasize:

Shape information, via one or several projective 2-D contours

Surface color or reflectance texture, primarily on the surface or the skin, but possibly also in the eyes in order to gauge product freshness

The solution will involve a grid or mosaic or point detectors with wavelength interval coatings, and a periodic layout customized to the orientation and movement geometry.

If the sorting allows for N-class classification with template shapes/color textures, at given orientations, the layouts will be optimized accordingly.

For example, high speed gutter control involves layouts with stacked triangles (Fig. 1), with apertures related to size and species, and median lines aligned with the knives.

Another example is weight estimation of live moving fish in aquaculture, where the spacing between detectors will be quantized in length and thickness increments corresponding to given weight increments (Fig. 2).

A third example is salt fish product grading by shape (especially elongation ratio (L/l)) and by reddish color tones (mapped in the proper color discrimination space by selecting properly the cut-off wavelengths of the coatings). In this case, the coatings will alternate with noncoated detectors, and the geometry will be radial to achieve invariance with respect to rotation (Fig. 3). It should be noted that the multispectral setup can involve two sensing arrays starring at the same field of view, or a mirror/beam splitter setup.

Basic related design or implementation techniques are found in Refs. 1 and 2, including vision classification or discrimination feature statistics. Non–vision detector outputs, suitably memory mapped for registration and movement compensation, can yield normal 2-D gray-level images subsampled at the detector positions only. For example, on can unfold at a sequence of contiguous instants the detector outputs acquired concurrently.

Figure 1 Gutter control.

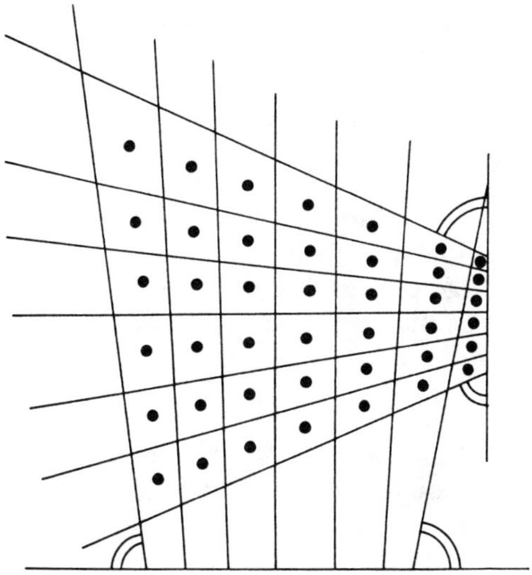

Figure 2 Weight estimation of live moving fish.

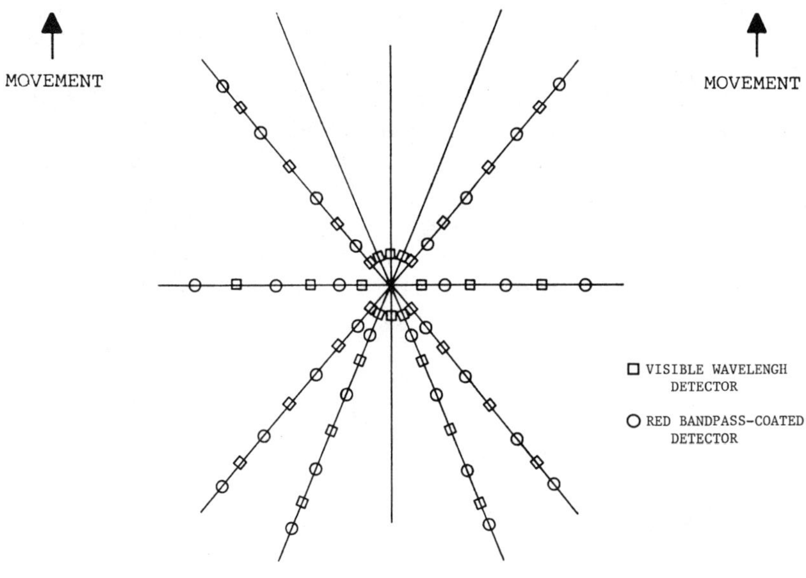

Figure 3 Saltfish product grating.

3. SENSOR FUSION FOR FISH PRODUCT INSPECTION AND HANDLING

Second-generation fish or fish product inspection and handling systems will emphasize:

Fusion/correlation of different sensing principles to detect and locate reliable specific defects or bones

Morphological image processing (e.g., skeleton) to generate the internal fish structure (meat/bones/etc.) from external shape(s) and defect positions

For example, feeding filleting machines will involve combined video/radioscopic images of the fish to locate the bones from the X-ray channel, the length from the video, and the weight and orientation from both (Fig. 4).

Trimming fish cubes for irregular asperities will involve cube position and orientation registration from video or CCD and precise following of the nominal contour by a laser beam to pick up gaps or too big spillovers of meat in three dimensions.

Rough volume and shape control on fillets will involve imaging by video or linescan in two perpendicular image planes to correlate shape with thickness dynamically.

Sensor fusion involves a number of basic steps, such as positional registration, attribute aggregation, and attribute fusion, which furthermore allows for concurrent processing and high information compression rates (3).

The morphological image processing methods needed above require simultaneously iterative connected components analysis (or image segmentation) and iterative morphological

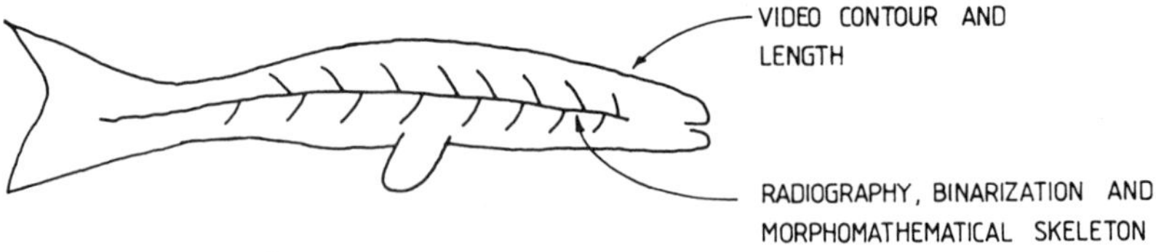

Figure 4 Video/X-ray sensor fusion with skeleton of bone images for filleting control and possibly species determination

feature extraction. This is still not readily known in the literature (4).

4. COLOR SPECTRUM ILLUMINATION FOR QUALITY CONTROL AND FRESHNESS GRADING

Whereas lighting for fish vision applications is quite difficult (wet curved surfaces, stained transport belt backgrounds, fish vessel environments), it is remarkable that humans operate quite consistently (but subjectively) when assessing quality and freshness.

It is here conjectured that the fish, fish product, and fish organ's color patterns provide good grading cues for those tasks. But the vision sensor spectral sensitivity and light source emission spectra as encountered in first-generation systems are still so mismatched to the optical properties of the fish that color pattern cues cannot be extracted.

Future systems will incorporate static or sweeping color modulators in the lighting while keeping constant the luminance. Thanks to this, the color patterns on the fish will be easily mapped with existing broadband vision detectors (video or CCD).

The reason is that the contrast variance will be improved by the said color spectrum illumination: Lighting a colored fish surface element with narrowband colored light of the same color will reduce it.

This can be implemented quite cheaply by sampling with a mirror or rotating prism the light from a special band in a color spectrum projected in a stationary fashion. The light of that color band can then be collimated onto all the fish, or the color spectrum itself is swept across the fish. Online operations only require a video frame rate high enough.

Such a color spectrum active lighting technique is ideal for grading of freshness, but also useful for fish species discrimination.

5. CONCLUSIONS

One might wonder why algorithms have not had a larger role in this chapter. Some promising vision algorithms classes are reviewed in Ref. 5, together with the limitations of conventional approaches. However, fish and fish product applications are most challenging at the feature extraction end emphasized here and elsewhere in this volume, and the required robustness imposes solutions that are not too dependent on algorithms only.

REFERENCES

1. Pau, L. F. and M. Y. El Nahas. 1981. An infra-red source classification system. Pattern recognition J. 14(1–6): 283–288.
2. Pau, L. F. 1980. Detector layout and scanning for an infra-red source classification system. SPIE Proceedings 253:160–166.
3. Pau, L. F. 1989. Knowledge representation approaches in sensor fusion. Automatica (2):207–214.
4. Giardina, C. R., and E. R. Dougherty. 1988. *Morphological methods in image and signal processing.* Prentice Hall, Englewood Cliffs, NJ.
5. Pau, L. F. 1984. Approaches to industrial image processing and their limitations. J. Electronics and power (IEE, London):135– 141.

Appendix

Users of Vision Technology in the Fishing and Fish Processing Industries

Fish Technology Research Institute, Tromsoe, Norway (research)

Frionor Norsk Frossenfisk, Oslo, Norway

Soelumithstoeth Hrathfrystihusanna, Reykjavik, Iceland

Sjavarafurthardeild Sambandsins, Reykjavik, Iceland

SIF, Reykjavik, Iceland

Samlecentralen Hanstholm, Hanstholm, Denmark

Hugo Moellebjerg, Esbjerg, Denmark

Findus (Nestle group), sites: Hammerfest, Norway; Bijuv, Sweden (research)

Denmark Fisheries and Oceanographic Research, Charlottenlund, Copenhagen, Denmark (research)

AI & Vision Group, Technical University of Denmark, Bldg. 348, Lyngby, Denmark (research)

Technological Institute, Taastrup, Denmark (research)

Danish Aquaculture Institute, Hoersholm, Denmark (research)

IFREMER, Brest, France (research)

University of Montpellier, Montpellier, France (research)

Peche et Froid, Lorient, France

MACSEA, Bayonne, France

Taiyo Fishery, Yokosuka, Japan

Nippon Suisan Kaisha Ltd., Hachiojii and other sites, Japan

Kyokuyo, Hachinobe, Japan

Nichiro Gyogyo, Kurihama and other sites, Japan

Hoko Fishing, Osaka, Japan

Hohsui Corp., Tokyo, Japan

Center of Marine Disease Control, Battelie Institute, Sequim, Washington, United States

Canadian Institute of Fisheries Technology, Halifax, Canada (research)

PISCES Computing, Vancouver, British Columbia, Canada (research)

Universal Multifood Ltd., Vermeuse, Newfoundland, Canada

Torry Research Station, Aberdeen, Scotland

IVP Department, TNO CIVO Institutes, I. J. Muiden, Netherlands (research)

RIVO Royal Institute for Marine Biology, I. J. Muiden, Netherlands (research)

EWOS (Alfa Laval Aquaculture Subs.) Soedertalje, Sweden

Fishing Industry Research Institute (FIRI), Cape Town, Republic of South Africa

Grove Telecommunications, Nova Scotia, Canada

Can Polar, Newfoundland, Canada

Tavel, Nova Scotia, Canada

Fisheries Production, Newfoundland, Canada

Index

Aniskids, 79, 89–90
Appearance, 5, 14
Aquaculture (*see* Fish farms)
Area, 253
Area camera, 175–178
Attenuation coefficient, 52
Averaging, 146–148

Backlighting, 31–32, 101–106, 127
Beam splitter, 109
Binary image, 230–233, 270–277
Black skin, 74
Bleeding, 7, 86
Blood spots, 17, 37, 74, 125–144, 233–236
Blurr, 257–258
Bone, 17, 26, 34–35, 213, 288
Bright field, 116
Calibration, 168–169
Camera geometry, 65
Candling tables, 35, 48, 84, 86
Catch area, 5, 15, 196–197
Chill store, 10–11
Chilled seawater, 7
Collagen, 35
Color, 16, 26, 74–75, 112, 285, 289
Complexity, 248
Condition factor, 9
Confusion matrix, 124
Containers, 269
Conveyor, 30–31, 172, 255
Counting, 265–280
Cubes, 25

Dark field illumination, 109
Defects, 17, 26, 34–38
Detection probability, 125

Diffuse reflection, 60
Diffuse transmission, 51–53

Edges, 239–240
Elongation, 286
Environment, 28–29
Epidermis, 73
Error probability, 125
Eyes, 285

Farnesyl methyl ether (FME), 80
Feature discrimination, 124
Fiber optics, 164
Filleting machines, 10, 288
Filters, 32, 140–143, 145, 237–240
  high-pass, 239–240
  laplacian, 239–240
  low-pass, 239–240
Fish farms, 251, 263–280
Flesh, 27–28, 49, 54–58, 74–75, 81, 105, 133, 149, 214
Flow injection, 17–18
Fluorescence, 16–17, 35–36, 61–63
Fourier, 157–158
Freshness, 16, 285

Glinting, 74, 126
Grading, 6, 9–10, 286–287
Grating, 155–156
Gutting, 7, 33, 173, 286

Handling, 6–9
  layout, 7–8
Height, 114, 117
Hormones, 84

Illumination, 31, 39, 97–119, 123–144, 161–166, 237–239, 251, 253, 269–270
Image acquisition, 100, 211–212, 227–228
Integrating sphere, 52, 102, 109, 129
Internal defects (*see* Defects)

Label, 5, 8, 24
Lamps, 105, 161–163
Label, 5, 8, 24
Lamps, 105, 161–163
Larvae, 79–80, 83, 189, 195, 268–270
Laser scanning, 66–67, 114–115, 163–167, 288
Length, 8–10, 24, 105, 114, 215–218, 249, 252–261, 288
Lidar, 158–159
Line scan, 115, 172–175
Light-emitting diodes (LED), 10, 23, 31, 117, 249
Light sectioning, 160, 162
Lighting (*see* Illumination)
Light sources (*see* Lamps)
Liver, 26
Look-up-table (LUT), 266

Moiré fringes, 155
Moiré topography, 155–158
Molting, 80
Moments, 217
Morphology, 219, 254, 258, 288
Mucus, 73

Nematodes, 47–68, 79–91
Neutron radiography, 36
Non-vision sensors, 283–290

Odor, 5, 14, 84
Optical properties of fish, 26–27, 71–75
Optics, 270–271
Orientation, 214, 288

Parasites (*see* Worms)
Polarization, 74
Polarized lighting, 106–109
Portioning, 10, 12
Price, 6, 8
Production, 6
Pseudo-color, 201–202

Quality assurance, 3–20

Refrigerated sea water, 7
Region detection, 38
Remote sensing, 187–203
Roundworms (*see* Nematodes)

Salting, 85
Satellite, 187–203
Scales, 10, 24, 27, 73–75, 249
Scattering, 56–59, 68
Seals, 84
Seawater tanks, 85
Sensor fusion, 288–289
Sensors, 13, 28–30, 283–290
Sex, 212
Shape, 25, 217–218, 285–288
Shelf life, 11, 14
Ship-to-shore communications, 198–203
Skeleton, 288
Skin, 27, 37, 73–74, 125–144, 233–237, 285
Sorting, 7, 9–10, 24–25. 30–34, 213–214, 247–261
Species, 10, 14, 24, 67, 73–74, 188, 192, 195, 213, 215–218, 252
Speckle, 169–170
Spectral absorption, 49–55
Spectral reflectivity, 134–136
Spectral transmission, 31, 49–50, 286
  flesh, 52–54
Spoilage, 5
Squid, 83
Stroboscope, 105, 133
Structured lighting, 113–117, 159–169
Surface defects (*see* Defects)

Tasmanian Food Research Unit System (TRFU), 14–16
Temperature, 5, 7–8, 11, 85, 187, 191
Texture, 5, 16, 26, 285
Thickness, 253
Three-dimensional vision (3D), 151–181
Threshold, 172, 231–232
Time, 5, 7–8, 11
Topographic vision, (2 1/2 D), 153
Total volatile base, (TVB), 12–13, 17–18

Trimethylene (TMA), 12–13, 17–18
Trimming, 10, 288

Ultrasound, 35

Visibility, 58, 61, 64
Vision, 23–24, 225–244, 285

Water control, 26, 31
Water jet, 33, 250
Weight, 5, 8–9, 24–25, 253, 286–288 (*see also* Scales)
Whales, 85
Worm detection, 25–26, 34–35, 48–68
Worms, 79–91, 213

X-ray, 35–36, 48, 215, 288

Yield, 9–10, 25